Gaston Boissier, D. Havelock Fisher

Rome and Pompeii

Archaeological Ramble

Gaston Boissier, D. Havelock Fisher

Rome and Pompeii
Archaeological Ramble

ISBN/EAN: 9783744782722

Printed in Europe, USA, Canada, Australia, Japan

Cover: Foto ©berggeist007 / pixelio.de

More available books at **www.hansebooks.com**

ROME AND POMPEII

UNIFORM WITH THIS VOLUME.

THE RIVIERA: Ancient and Modern. By CHARLES LENTHÉRIC. Translated by CHARLES WEST, M.D. With Maps and Plans.

ROME AND POMPEII

Archæological Rambles

BY

GASTON BOISSIER
OF THE FRENCH ACADEMY

TRANSLATED BY

D. HAVELOCK FISHER

WITH MAPS AND PLANS

NEW YORK
G. P. PUTNAM'S SONS
27 AND 29 WEST 23D ST.
1896

CONTENTS.

CHAPTER I.

THE FORUM 1-50

I. Importance of the Forum down to the end of the Empire—Its condition at the beginning of the present century—Signor Pietro Rosa's excavations—M. Dutert's restoration essay—Signor Fiorelli's administration, ... 8-14

II. *The Via Sacra* between the Arch of Titus and the Forum—The Temple of Vesta—The dwelling of the Vestals (*Atrium Vestæ*)—The Vestals and the Christian nuns—View of the Palatine, . . . 15-40

III. *The Forum of the Empire*—How we have been enabled to recognise and designate its chief monuments—Statius and the statue of Domitian—The Temple of Cæsar—The *Basilica Julia*—Temples of Saturn and Castor — Those of Vespasian and Concord—East side of the Forum—Centre of the Forum—The *Clivus capitolinus* 41-53

IV. Impression first produced by the Forum—Absence of symmetry—Its small extent—The very different uses it served—Political assemblies—How orators made themselves heard in it—How it held all the people who came together there, . . . 53-66

CHAPTER II.

THE PALATINE 67-137

I. How the excavations on the Palatine came to be undertaken—*Roma quadrata* and the walls of Romulus—The Temple of Jupiter Stator—Remains of the epoch of the Kings—Antiquity of writing among the Romans, and the consequences to be drawn from it—The Palatine under the Republic—Why excavations are always so prolific in Rome, . 71-86

II. The house of Augustus on the Palatine—How, little by little, it became a palace—What remains of it—Employment of marble in the Imperial epoch—New processes in the art of building—The Palace of Tiberius—That of Caligula—The *cryptoporticus* where Caligula perished—The house of Livia and its paintings—The Palace of Nero, ... 87-111

III. The *Flavii* and their policy—Description of Domitian's palace—The palace of Severus—The Imperial box at the Great Circus—Lodgings of the soldiers and servants, ... 111-129

IV. Aspect of the hill in the third century—It contains the edifices of all times—Monuments of the Imperial epoch—Differences between the palaces of then and now—Beauty of the whole, . . 129-137

CHAPTER III.

The Catacombs 139-213

I. The importance which Christians attached to sepulture—The Catacombs their work, and not old abandoned quarries—How they were induced to hollow them out—Hypogea of different religions in the Roman Campagna—Rules adopted by the Church for burial, ... 142-152

II. First impression produced by a visit to the Catacombs—The immensity of these cities of the dead, and consequences to be drawn from it—Rapid diffusion of Christianity—Religion separates itself from the family and the country—The Catacombs the most ancient monument of Christianity at Rome —Mementoes of the times of persecution contained in them—Mementoes of the days of triumph, 152-161

III. The inscriptions and paintings in the Catacombs—Character of the most ancient inscriptions—The birth of Christian art—The first subjects treated by the artists of the Catacombs— Imitation of antique types — Reproduction of Christian subjects—Symbolism—Origin of historical painting—To what extent the Christian artists adhered to antique art, ... 162-177

IV. The cemetery of Calixtus—Signor Rossi succeeds in finding it—The indications which enable him to discover the tombs of the martyrs—Works carried out after the time of Constantine in the celebrated crypts—*Graffiti* of pilgrims—Why the cemetery took the name of Calixtus—History of this Pope, according to the *Philosophumena*—Why the Popes of the third century were buried in the cemetery of Calixtus, and how it became the property of the Church—Discovery of the papal crypt, 177-192

V. Chief results of Signor Rossi's discoveries—
His new opinions on the origin and history of the
Christian cemeteries—They begin by being private
property—As such they are under the protection of
the law—How they extended—How they became
the property of the Church—First relations of the
Church with the civil authority—Character of these
relations—The primitive Church and the great
families—How advantage may be drawn from the
acts of the martyrs, ... 192-213

CHAPTER IV.

HADRIAN'S VILLA 215-292

I. The Emperor Hadrian—The different judgments passed on him—The prince and the man—
The reasons why he was not loved—His liking for
the Greeks — Travelling in ancient times —
Hadrian's journeyings, ... 221-240

II. The site of Hadrian's villa—Magnificence of
construction—The Emperor's purpose in building
it—Parts which can be recognised—The Vale of
Tempe — The Pœcile — Canopus -- The private
dwelling—The Natatorium—The reception apartments—The *Piazza d'oro*—The Basilica—The
theatres—The libraries—The public lecture-halls—
Hell 241-268

III. Did the Romans understand and love Nature?
—The reasons they had for leaving the town—
Horace at Tibur—Liking of everybody for the
country—How Pliny the Younger lived there—His
villas — His gardens — Sites preferred by the
ancients—The view from the Pœcile, . . 268-292

CHAPTER V.

OSTIA 293-334

I. Modern Ostia—Aspect of the plain by which ancient Ostia is covered—How the town came to be abandoned — The first excavations made there— Signor Visconti's labours—Discovery of the Street of Tombs—The house known as the Imperial Palace—The great temple and the street leading towards the Tiber—The shops situated along the river, ... 295-306

II. Why the port of Ostia was founded — The free distribution of corn in Rome—The difficulty of provisioning Rome—Creation of the port of Claudius — The port of Trajan — The Imperial Palace—The town of *Portus*—The magnificence of *Ostia* and *Portus*, ... 306-325

III. The religious monuments at Ostia—Introduction and swift progress of Christianity—The *Xenodochium* of Pammachius — Prelude to the *Octavius* of Minutius Felix—Death of St Monica, 325-334

CHAPTER VI.

POMPEII 335-485

I. The excavations at Pompeii under Signor Fiorelli—Mementoes of its ancient history that have been found—What remains to be cleared—Ought the works that have been begun to be continued? —Recent discoveries—The fresco of Orpheus— Account-books of the Banker Jucundus—The new *Fullonica*, ... 335-354

II. Pompeii's chief lesson to us—Country life in the Roman Empire—The difficulty of acquainting ourselves with it—How Pompeii puts it before our eyes—The whole Empire repeats the customs of Rome—The aristocracy of Pompeii—Characteristics of Pompeian houses, ... 354-369

III. The paintings of Pompeii, according to Doctor Helbig's works—The large number of mythological pictures—Character of these pictures—The paintings of Pompeii not original—Why critics of the first century treat the paintings of their time so severely—From what schools did Pompeian artists borrow the subjects of their pictures ?—Alexandrian or Hellenistic painting—Room pictures—General character of Hellenistic painting—How far did Pompeian artists faithfully reproduce their models ?—What is the particular merit of the paintings at Pompeii ?. . . 370-398

IV. Whence the resemblances come that are remarked between the paintings at Pompeii and the poetry of the Augustan age—The painters and the poets inspired by the same subjects—Latin literature imitates the poetic school of Alexandria—Catullus — Virgil — Propertius—Ovid—Differences between the painters of Pompeii and the Roman poets—Painting never became Roman— Repugnance of the Pompeian artists to handle subjects drawn from the history or the legends of Rome — Is Pompeii really a Greek town ?—National character of the poetry of the Augustan age, . . . 399-419

V. The burghers of Pompeii—The poor—Where did they live?—Inns and taverns—Occupations and pleasures common to the poor and the rich—The municipal elections—The shows—How may we become acquainted with the inner life of the Pompeians?—The inscriptions and *graffiti*—The services they render us, ... 419-435

ARCHÆOLOGICAL RAMBLES

CHAPTER I.

THE FORUM.

I HAVE often heard it said that it is dangerous to return after long absence to persons and places one has much loved. We seldom find them again as we remember to have seen them. The charm flies with years, tastes and ideas change, the faculty of admiring wanes; there is a danger of our remaining unmoved before what transported us in our youth, and, it may be, that instead of the pleasure we sought, only a disappointment awaits us. This disenchantment is the more fatal in that it usually spreads from the present to the past. Do what we will, it ends by imparting itself to our old impressions, and taints those stores of memories which should be faithfully treasured in our hearts for life's decline.

And this is the peril to which a traveller exposes himself who, not having seen Rome for many years, determines to go back there. How many things have happened in these few years! Rome has changed masters; the old town of the Popes has become the capital of the kingdom of Italy. How has she lent herself to the change? What effect has this new order of things, so different from the old, produced upon her? Has she lost anything by it, and shall we find her again as we left her? This is the first question we ask ourselves on returning to Rome. It is difficult not to feel engrossed by it; and directly the railway lands us upon the immense piazza of the Baths of Diocletian—once so calm, now so bustling and noisy — we cannot help looking round on all sides with uneasy curiosity.

The first impression is not favourable, it must be owned. On leaving the terminus, we traverse a new quarter, which offends by its likeness to every other new quarter in the world. Is Rome, then, in peril of becoming a commonplace town? We see vulgarly elegant houses, like those we have seen in other cities; we pass an immense building—a species of barracks, without character or style, destined to become a public office, and which produces a pitiful effect beside the grand palaces of the sixteenth century; and, as we go through broad streets and narrow lanes flooded with a burning sun, we remember that, even in the time of Nero, who rebuilt the old town on a vaster plan, boobies much admired the splendour of the new building, but sensible people could not help regretting the old narrow, crooked streets, where they always found so much shade and freshness.[1] This is hardly an encouraging beginning, and what remains seems at first in keeping with it. On descending from the Quirinal to the *Corso*, we still find many striking changes. The *Corso*, with the streets that

---

[1] Tacitus, *Ann.*, XV. 43: *Erant tamen qui crederent veterem illam formam salubritati magis conduxisse.*

cross it, from the *piazza di Venezia* to the *piazza del Popolo*, was always the most animated place in the town. It appears to me to have become still more animated, and that its population is no longer quite the same. Priests, and especially monks, are more rare, and the glance of those who remain does not seem to me so assured, nor their countenances so proud; they evidently no longer feel themselves the masters. Among the people who have replaced them, one is surprised to see many who walk fast, and appear to have something to do, which used to be seldom the case. Nor do they belong to the old inhabitants of Rome. They are generally employés of the ministries, or public office clerks; all come from outside, bringing with them new customs. At the very hour when, according to the old saying, only dogs and Englishmen were seen in the streets, we meet these officials, active, busy, elbowing those who are in their way, to the intense amazement of the old Romans, who cannot understand people going out at the hour of the *siesta*, or hurrying when it is hot. As evening approaches, the bustle increases. There is a moment, towards six o'clock, when the street belongs to the news-vendors. They deafen you with their cries; they address you, they pursue you. Newspapers abound in Rome. There are journals of every size and shade of opinion—more violent than moderate, as usual—which bid for clients by the smallness of their price and the vivacity of their polemics. How far are we from the time when only that good, carefully expurgated *Giornale di Roma* was read — that friend of legitimate governments, which never knew of revolutions until several weeks after they had taken place! Must we believe that

this race of sceptics and scoffers, accustomed and
indifferent to everything, astonished and indignant at
nothing, which used to answer the reformers of all
parties with a *che volete?* or a *chi lo sa?* has suddenly
gone raving mad over politics? It is a change one has
great difficulty in understanding. And it is impossible
to master one's surprise when the very signboards are
seen to contain professions of faith—the barbers pompously styling themselves *parruchiere nazionale*—and
when one reads the electoral appeals and the demo-
cratic bombast with which the walls are covered. There
are certainly many innovations which run great risk of
not being to the taste of everybody. We cannot help
asking ourselves what will be thought and said by those
zealous admirers, whom Rome has possessed in all ages,
who would have her remain stationary; who say she is
being spoilt when the least thing is changed; and who
already began to cry out that all was lost when a too
zealous magistrate took it into his head to have the
streets a little better swept, and to light them dimly
with a few lamps.

But let us hasten to reassure them. All is not so much
overturned as they may think, and the change is more
upon the surface than in the depths. The quarters of
the people have nearly everywhere kept their old aspect.
If, for example, after descending the *Corso*, you continue
your walk beyond the *piazza di Venezia*, through the
steep streets leading to the Forum, you find old Rome
intact. These, indeed, are the same houses we used to
see—as old and as dirty. The Madonnas have remained in their places above the doors, and in the
evening a lantern is still piously lit before them. If

you happen to raise your eyes higher, towards the wide, curtainless windows, you are sure to find enough rags spread out to content the most exigent friends of the picturesque and of local colour. In the cellar-like taverns, with their great open doors, players are still leaning carelessly with their elbows on the tables, beside flasks of *Orvieto*, with greasy cards in their hands. As for the *osteric* which skirt the street, I do not think they can have much changed in appearance since the Roman Empire; and I muse, as I behold them, on those *unctæ popinæ* whose rejoicing smell so gladdened the slave of Horace.

So, with a little good-will, here we are in the very midst of antiquity. If we wish the illusion still more complete, if we desire to enjoy for a moment what might be called the genuine sensation of Rome—that which our fathers felt in visiting it, and which was described by Chateaubriand and Goethe—let us go a little further, beyond the houses and the boundary: for, in order to insure a better understanding of it, it is as well to leave them behind us. If you like, we will pass through the Porta Pia, and follow the ancient *Via Nomentata*. Saluting, as we go by, the basilica of St Agnes and the round temple that served for the sepulture of Constantine's daughter, we get to the Teverone, and cross it by a very curious bridge, still bearing traces of work dating from the Middle Ages. A few steps further on, to the right, rises a hill of small extent and height. It must be climbed with respect, since it bears a great name in history: it is the Holy Mountain. Here it was that, more than two thousand years gone·by, Democracy gained one of its first victories using, in order to obtain it, a means it is still very fond of employing—the *strike*. One fine day, the Roman army—that is to say, all the sound-bodied population—leaving the camps to which the Consuls insisted on confining it, came and settled on this mountain, determined to remain there so long as its conditions should be refused. In order to win, it only had to wait. The Aristocracy, alarmed at its solitude, became weary of resistance, and allowed the people to institute the tribuneship. How many memories present themselves to the mind from the summit of this hill! It was in this immense undulating plain which now strikes the eye that, according to the expression of an historian, the Romans served their apprenticeship for the conquest of the world. Every year they had to fight the energetic little tribes peopling it, and furious battles took place there for the possession of a hovel or the sacking of a cornfield. There it was that, during a struggle of many centuries, they acquired experience of war, the habit of obedience, and ability to command. When they crossed those mountains which frame in the horizon on all sides, in order to spread themselves over the rest of Italy, their education was completed, and they possessed the virtues which enabled them to conquer all. Since then, how many glorious events! Since then, how many times have those great roads, whose direction is still followed by the line of tombs that border them, witnessed the return of the triumphant legions! How many illustrious names are recalled to the memory by those fragments of aqueducts, and those ruined monuments which cover the plain! And we have here the advantage that, these great memories once revived, there is nothing to divert us from them. In fertile, well-peopled countries, full of bustle and movement, the present unceasingly snatches us away from the past. How can we muse and ponder, when the spectacle of human activity craves our attention at every moment, and from all sides the noises of life reach our ears? Here, on the contrary, all is silence and contemplation. As far as the eye can range, nothing is seen but a naked plain, sparsely covered with thin grass, without trees, except some scattered parasol pines, and, beyond a few taverns for sportsmen, devoid of houses. The landscape only strikes as a whole. It is a general monotony, or rather harmony, where everything melts and blends. Nothing draws the attention; no detail stands out with undue prominence, nor jars. I know no spot on earth where one can allow one's thoughts to carry one away more completely, and absolutely give Time the slip; as Titus Livius so aptly expresses it: "Where it is easier for the soul to become antique and contemporaneous with the monuments it gazes upon." The Roman Campagna has kept this advantage in perfection, nor is it easy to foresee when it will lose it. Many projects are made to render it healthy and people it, but Death has entered so deeply into this exhausted soil, that he will probably not be dispossessed without trouble. In the meantime, let us enjoy the privilege which this country preserves of putting us, better than any other, in communication with the past. Whatever effort Rome may make to adorn and embellish herself, and be on a level with the fashion of the day, it is

Antiquity one goes there to seek above all things, and, thank God! it is still to be found. With those great ruins with which she is strewn, and the vast deserted plains that surround her, she has not, nor will she, for a long time to come, be able to give herself as modern an air as she would desire. That she should have succeeded so little is happy for her and for us; for we may apply to her what a poet of the Renaissance said of Michael Angelo's "Night"—"'Tis because she is dead she lives" (*perche ha vita !*).

I.

IMPORTANCE OF THE FORUM DOWN TO THE END OF THE EMPIRE—ITS CONDITION AT THE BEGINNING OF THE PRESENT CENTURY—SIGNOR PIETRO ROSA'S EXCAVATIONS—M. DUTERT'S RESTORATION ESSAY.

EVERYTHING, in fact, invites people who visit Rome to busy themselves chiefly with Antiquity; for up to the present moment it is Antiquity which seems to have profited most by the events of 1870. The new government owed much to ancient memories; since a favourite expedient for emphasizing the right of Rome to be free and dispose of herself, and of Italy to claim the city as a capital, had been to appeal to the history of the Republic and the Empire, and to talk unceasingly of the Senate, the Forum, and the Capitol, the new pretensions gaining considerably by the protection of these great names. The Italian Government had thus contracted a debt with the past, which it set about paying as soon as it was installed in Rome. As early as the 8th November 1870, a decree of the king's *locum tenens* instituted a superintendence of excavations, and charged therewith the skilful explorer of the Palatine, Signor Pietro Rosa. A week later the works in the Forum began.

It is natural that attention should have first been turned in this direction. The Forum enjoyed the rare good fortune of remaining in all times the centre and heart of Rome. In nearly all our modern capitals the focus of activity and life changes with the centuries. In Paris it has passed successively from the left bank of the Seine to the right, and from one end of the town to the other. Rome proved more faithful to her ancient traditions. From the day when, according to Denys of Halicarnassus, Romulus and Tatius, established the one on the Palatine and the Cœlian, the other on the Capitoline and the Quirinal, decided to meet, for the discussion of common affairs, in the damp unwholesome plain stretching from the Capitoline to the Palatine,[1] it never ceased to be the city's place of meeting and council. During the first years there was no other public place, and it served for every use. In the early morning all kinds of goods were sold there, throughout the day it was a court of justice, and, in the evening, people took their walks there. As time went on, public places multiplied, and there were special markets for cattle, for vegetables, and for fish (*forum boarium, olitorium, piscatorium*); but the old Forum of Romulus always retained its pre-eminence over all the others. Even the Empire, while changing so many things, did not deprive it of this privilege. Public places were

---

[1] Denys, II. 50.

built round about it, more vast, more regular, more sumptuous, but which were never otherwise looked upon than as the annexes and dependencies of what people persisted in calling the real Roman Forum. It held out against the first disasters of the invasions, and survived the taking of Rome by the Visigoths and the Vandals. After each storm, the Romans set about repairing it as best they could, and even the barbarians themselves, as in the case of Theodoric, sometimes took the trouble to restore the buildings they had ruined. The old place and its buildings still existed at the beginning of the seventh century, when it unhappily occurred to the Senate to consecrate to the abominable tyrant Phocas that column of which Gregorovius tells us, "the Nemesis of history has preserved it as a last monument of the baseness of the Romans." From that moment ruins accumulate. Each war, each invasion, throws down some ancient monuments, and no trouble is taken to repair them. Temples, triumphal arches, that have been flanked with towers and crowned with battlements, like fortresses, attacked every day in the struggle of parties which divide Rome, and shattered by assaults, end by falling, and cover the soil with their ruins. Every century adds to this accumulation. When, in 1536, Charles the Fifth went through Rome on his way back from his expedition to Tunis, the Pope wished to make the avenger of Christianity pass beneath the Arches of Constantine, of Titus, and of Severus, and nothing was spared in order to provide him a finer road. "They demolished and pulled down more than two hundred houses and razed three or four churches, level with the ground," says Rabelais, who witnessed it. It is said that a few years later, Sixtus V. had the débris of the building materials, which he was using elsewhere, transported to this desert spot. All antiquity found itself covered over and lost beneath more than 6 mètres of rubbish. From that moment the Forum, now the *Campo Vacchino*, or cattle field, assumed the aspect which it kept until the beginning of this century. It was now only a dusty, open space, surrounded by mediocre churches, about which a few columns rose, half protected by the soil, a melancholy and forlorn spot, quite suited for reveries on the frailty of human grandeur and the vicissitudes of events. This is how Poussin represents it in his little picture in the Doria Gallery, and Claude Lorrain does the same in the landscape at the Louvre.

One would think that these half-buried columns would have provoked the curiosity of the learned. How happens it, that since the Renaissance not one of them has undertaken to excavate to their bases, in order to discover the soil they rest on? This soil was that of the Forum; it was known beyond a doubt that it would be found strewn with historical ruins; and yet no thought was ever seriously entertained of undertaking works which might lead to the finest discoveries. It was only in the first years of this century that learned researches began; but they were too often interrupted, and gave rise to more problems than they solved. The information they elicited was so incomplete, that fierce contests arose between the archæologists. Each gave a different name to the buildings that were brought to light, and each made for himself a special plan of the Forum. Neither its exact limits nor even its precise position were known. Some supposed it must extend from the Arch of Severus to that of Titus—that is to say, from north to south [1]— while others placed it in quite the contrary direction, viz. from St Adrian to St Theodore; all believed they found in the ancient writers texts clearly confirmatory of their opinions. In order that this confusion might be dissipated, fresh excavations were indispensable. They were undertaken with the idea of this time carrying out a work which should be definite. It no longer sufficed to try a few soundings to touch the ancient soil here and there; it was resolved to free it entirely from the rubbish that covered it, and lay it bare in every part. This was the means adopted finally to ascertain the truth respecting the enigmas of the Forum.

Signor Rosa first resumed the excavation of the Julian basilica, which had been partially cleared under the late Government. This work ended, the whole of one side of the Forum was known and acquired, namely, that extending to the west from the slope of the Capitoline to the first spurs of the Palatine. The workmen were urged forward towards the east, and no stop was made until the churches of Santa Martina and St Adrian were reached. The municipal Council of Rome would not permit a farther advance, being

---

[1] Although these designations are not quite exact, I call north side of the Forum that situated at the foot of the Capitoline, and south side, that which extends from St Lorenzo in Miranda (Temple of Antoninus) to St Maria Liberatrice. The east side is that bordering the churches of Sta Martina and St Adrian; the west side, that stretching from the Via della Consolazione to the Palatine.

unwilling to allow the destruction of the streets joining the different quarters of the modern town. However vexing this check, one had to be content with what it had been possible to do. In justice to Signor Rosa it must be owned that the works directed by him were vigorously prosecuted. It was necessary to remove more than 120,000 cubic mètres of earth, but, under it, many ancient monuments were found which were only known by name, and at several points the topography of the Forum has been fixed. It is to be regretted that the Roman administration did not deem it necessary to publish a detailed journal of these interesting excavations; but this gap has happily been filled in part by the work which a young member of the French school at Rome, M. Ferdinand Dutert, has published on the Forum, and of which I am about to make liberal use.[1] M. Dutert assisted at Signor Rosa's labours and followed their progress day by day, walking behind the workmen, gathering and copying the least remains of ornaments and smallest fragments of sculpture as they met with them on their way. His work not only shows the present state of the Forum to those who have not seen it, and recalls it to those who have, but he has tried to teach us what it was in ancient times. He restores the ruined temples he raises again the fallen columns, he replaces the statues upon their bases, and puts once more before our eyes those splendours of which but a few fragments are left. I know there is always much conjecture in works of this kind; but

---

[1] *Le Forum romain* par M. Ferd. Dutert, architece, ancien pensionnaire de France à Rome, Paris, chez A. Levy.

M. Dutert's restoration, usually based on exact indications, is in general very probable. Only a few deficiencies and errors have been noted in it, which, in the present state of our knowledge, it was very difficult to avoid.

To ensure more activity, and, at the same time, more unity in these explorations, the Italian Government instituted a Direction-General of Antiquities and Fine Arts at Rome, and placed it under the charge of Signor Fiorelli, who had made a great name for himself by the able manner in which he had conducted the explorations at Pompeii. Signor Fiorelli from the very first made up his mind not to waste his energies and his resources on isolated excavations, but determined to concentrate his efforts on the Forum and its environs. The work had been well begun, and had produced the happiest results; the best thing to do was to follow it up. The large square lying between the Basilica of Constantine and the Palace of the Cæsars was yet to be explored. This vast space did not form part of the Forum proper, but it was the natural entrance to it, and was connected with it by the monuments with which it was crowded; so that it could not be set on one side. These explorations have taken ten years to accomplish; they are now complete. The ground from the Arch of Titus to the Capitol, a length of nearly 500 mètres, has been laid bare. Let us profit by this to explore it in its entirety, to study the buildings on it, and to awaken the memories of the past as we come across them on our way.[1]

---

[1] All the objects of our research may be studied on the *Plan of the Forum*, where they have been placed in their actual positions.

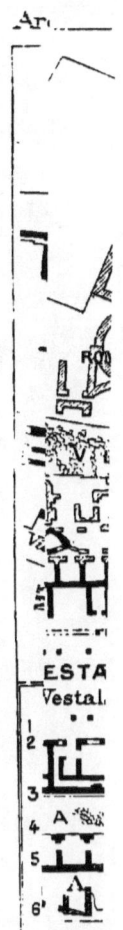

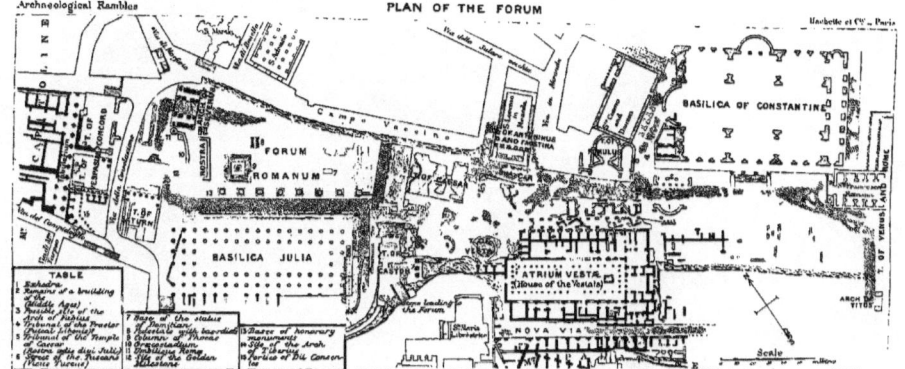

II.

THE *VIA SACRA* BETWEEN THE ARCH OF TITUS AND THE FORUM—THE TEMPLE OF VESTA—THE DWELLING OF THE VESTALS (*ATRIUM VESTÆ*)—THE VESTALS AND THE CHRISTIAN NUNS — VIEW OF THE PALATINE.

Visitors, as a rule, enter the Forum by the Temple of Castor, opposite the church of Santa Maria Liberatrice. Thus they find themselves at once in the very centre of the place. But for the better understanding of the arrangement of the Forum, I prefer to begin at the beginning and follow the road that used to be taken by the multitude. We will commence at the farthest extremity. I am supposing that we have just left the Colosseum, and that we are walking along the Palatine. We see stretched before us a wide ancient road, over the old flagstones of which the traffic of the modern town still rolls. This road rises straight before us over a fairly steep slope and under the Arch of Titus. We are on the *Via Sacra*.

The position of the *Via Sacra* has been the subject of many disputes among archæologists. We must not be surprised to find this question a difficult one to answer, for the ancients themselves do not seem to have been very clear upon this point. The example of Pompeii shows us that streets were not then inscribed with their names, and, as the knowledge of these appellations only became very gradually known, there was often much uncertainty concerning them. It was on this account that Varro and Festus tell us that "the multitude were not very sure which road they ought to call the *Via Sacra*." They add, however, that everyone agreed to give that name to the road which led from the Temple of the Lares (near the Arch of Titus) to the Temple of Vesta. At the present day we know this road perfectly well; we are able to traverse it in its whole length, thanks to the efforts of the excavators.

On leaving the Arch of Titus, the road makes a sharp turn to the right, and follows the course of a large terrace, which is raised several steps above it. It was on this terrace that the Emperor Hadrian had built his Temple of Venus and his Temple of Rome, of which some very fine ruins still remain.[1] After passing the Church of Santa Francesca Romana, with its elegant clock tower, it turns to the left, close to the Basilica of Constantine, from which it is separated by some modern buildings; then it passes in front of the Temple of Romulus (Church of SS. Cosmo and Damiano). This edifice, built by Maxentius in memory of his son who died young, was half buried in ruins. These have all been cleared away, and the door has been restored to its place; of the four Cipolino marble columns which ornamented the sides of the façade, two have been set on their bases; in fact, the little temple has been restored to its primitive elegance. The monuments on the other side of the road are neither so important nor so well preserved. On a level with the ground several bases of statues have been found. The right of placing

---

[1] M. Laloux has published a restoration of this building in the *Mélanges d'Archæologie et d'histoire* of the school at Rome.

THE FORUM.

one's statue by the side of a road so much frequented by the public was doubless a great honour and one much sought after; it was a sure means of keeping oneself always before the populace, and ensured a greater chance of being remembered. By the side of these honorary pedestals the remains of an exhedron have been found; that is to say, one of those semi-circular benches such as have been found at Pompeii, on which loungers might sit and chat, or watch the passers-by.[1] A little above and behind this first row, of which so little remains, the excavators have brought to light the whole of an ancient district composed of houses closely crowded together. This quarter must have become very dilapidated even in those early days; under the basements of the most recently-built houses the foundations of older ones, running in a contrary direction, have been discovered. The incendiary fires, which were of such frequent occurrence at Rome, especially in the low-lying Forum, often totally changed its aspect. M. Jordan thinks that it must have been entirely reconstructed in the time of the Emperor Hadrian, when he built his Temples of Rome and Venus, and naturally wished them to be placed amid suitable surroundings, the better to set off his skill as an architect.

Instead of following this road as far as the point where it joins the Forum, let us turn to the left for a moment. We will cross this block of houses whose foundations have been brought to light, and proceed

---

[1] See No. 1 on plan.

B

towards the Palatine and the Church of Santa Maria Liberatrice. This place has played an important part in the history of ancient Rome. It was here that the first kings established the religious centre of Rome before Tarquin transferred it to the summit of the Capitol. The building of the Temple of Jupiter marks a new epoch in the religious life of the Romans. The period which had preceded it, and which is sometimes called the period of Numa, bore a very different stamp; then religious rites were simpler, and the sacred buildings less sumptuous; no statues had as yet been raised to the deities, and cakes made of salted flour were the only sacrifices. There still remained, in the days of the Empire, three monuments of this primitive age which time had respected, and which were situated close together. These were the Temple of Vesta, where burnt the eternal fire; the *Regia*, or the dwelling of the king, who, being both the spiritual head and the chief magistrate of the city, had to live in a central position; and, lastly, the *Atrium Vestæ*, where the vestals resided who assisted the king in his capacity of high priest, in the same way that, among private people, the daughters helped the father of the family in the service of the household gods. These are the three monuments which were being sought.[1]

---

[1] For an account of the discoveries which have been made on this side, the reader is referred to the work of Signor Lanciani, entitled the *Atrio di Vesta*, published in the *Notizie degli scavi* of 1883, and to that of M. Jordan in the *Bulletino dell' instituto di correspondenza archæologica*, of May 1884.

The first discovery, that of the Temple of Vesta, was made several years ago. After the *Basilica Julia* had been cleared, the workmen, while digging at a short distance on the further side of the Temple of Castor, came upon a small round basement completely in ruins. Although so humble in appearance, there were archæologists who did not hesitate to assert that these foundations must have supported the famous temple whose origin has been ascribed to Numa. At the time, this statement led to much discussion; but no one has dared to dispute it since the dwelling of the vestals has been found quite close to it. Time is not altogether to blame for the fact that the only remains of this old temple should be a heap of earth and a few scattered stones. Time is less skilful than man in bringing about the ruin of ancient monuments; and, among men, the most highly civilised are often those to be most feared. "The excavators of the sixteenth century," says Signor Lanciani, "have done more harm to the antiquities than all the barbarians of the Middle Ages." In 1549, some archæologists, seeking for statues and other precious objects, discovered the Temple of Vesta, the ruins of which had been fairly well preserved under heaps of rubbish, but they lost no time in bringing about its entire destruction. They carried away, for edifices which they were building, marble facings, friezes, columns, and even blocks of volcanic stone which were used for foundations; they made lime of the stones which they did not care to take away; then, their devastations completed, they covered up all that was left with earth. Happily, a scientist of that time, Pavinio, had made a sketch of the ruins. This sketch, with the help of one or two bas-reliefs, and a few coins on which the temple is depicted, gives us a slight idea of what it must have been. It is ridiculous to say that the monument of which the ruins were found in the sixteenth century was not the one built by Numa; it might have been rebuilt more than once in ten or eleven centuries; but Ovid says that, in rebuilding it, it was altered as little as possible, and great care was taken that its ancient appearance should be preserved.[1] It was a round building, surmounted by a small cupola covered with sheets of metal. The *savants* invented very learned reasons to account for the round shape which they insisted in giving it. "It is round," they said, "like the world, and the world must be represented in the shape of a globe, in the centre of which burns the fire which nourishes everything."

"*Vesta eadem est quæ terra; subest vigil ignis utrique.*"

These subtle explanations of the ancient philosophers have now been abandoned, and we do not ascribe any such refined ideas to the rude peasants who, six or seven centuries before our era, built the first Temple of Vesta. It is supposed that they constructed it on the plan of the houses in which they themselves lived. Probably they knew of no other way of building. This is why the monuments which date back to the foundation of Rome all bear so much resemblance to one another; for example, the little hut of Romulus on the

---

[1] Ovid. *Fast.*, VI. 267.

Palatine, which has been preserved with such respect; the Temple of the Penates on the heights of Velia; that of Hercules Victor in the *Forum boarium*, all reproduce the shape of the round cabins which were the first dwellings of the Italian people.[1] These ancient buildings were afterwards very often repaired, and every time they were repaired they were enriched. Ovid says that marble had taken the place of the interwoven rushes which had formed the walls, and that the thatched roof had become a dome of brass;[2] but, as I have just said, a sort of instinct of preservation, which is peculiar to this people, caused them to retain the original dimensions, the same external shape and general aspect, so that in the midst of the splendours of the Empire they seem to have preserved some souvenir and some image of remote antiquity.

The dwelling of the vestals is situated, as might be expected, quite close to the temple in which they ministered. If, in 1876, the excavations had been prosecuted a little further, it would soon have been discovered; but they were directed towards another spot, and it was only after having excavated the whole length of the *Via Sacra*, from the Basilica of Constantine to the Arch of Titus, that the restorers returned to the Temple of Vesta. A very few blows of the pick were enough to disclose the walls of the house of the vestals: thanks to the activity with which the works have been carried on, it has now been entirely laid bare. This was, with-

---

[1] See Helbig, *Bull. dell' instit.*, 1878-9.
[2] Ovid, *Fast.*, VI. 261.

out doubt, the greatest discovery that had been made for many years; and, if we accept the *Basilica Julia*, it was the most important monument as yet found on the Forum.

The entrance is by a small side-door of unimposing appearance, but, after having crossed several steps, a rectangular court 68 mètres long by 20 wide is reached. This court corresponds to the peristyle in ordinary houses, but is of an obsolete style. It was surrounded by vast porticoes, decorated with statues of the *vestales maximæ* (the presidents of the College). These statues were placed on pedestals bearing pompous inscriptions. Signor Lanciani supposes that, at the time when the edifice was intact, it must have contained a hundred of these statues, but time has marvellously diminished the number. Now we have the fragments of but eighteen, all more or less mutilated. The pedestals are a little better preserved. Some had already been obtained from the excavations of the sixteenth century;[1] the later explorations had produced about twenty more, of which some are in a perfect state of preservation. They bear inscriptions which teach us much. They show us what consideration the vestals enjoyed, and how much they were mixed up in public affairs. It was considered such an honour to belong to their college that Tiberius, to console the daughter of Fonteius Agrippa, who had not been elected, is believed to have given her a million sesterces.[2] The honour

---

[1] The inscriptions which were known before these last excavations have been collected in the *Corp. insc. lat.*, VI. 2127-2145.

[2] Tacit, *Ann.*, II. 86.

was reflected on the whole family, and among the statues of which the remains have been found in the *Atrium Vestæ*, several have been raised by relatives who were proud of having a vestal in the family. Sometimes they were set up by people who wished to evince their gratitude to one of the priestesses for a favour they had received, and the nature of the benefit shows us how far the vestals' power extended. We are surprised to see that they contributed to the nomination of the Emperor's librarian;[1] but there are cases in which their interference astonishes us even more. How did they manage to procure for someone the rank of military tribune? and what good office could they have rendered to those centurions appointed by their comrades to arrange at Rome the affairs of their legion?[2]

It is not astonishing that the gratitude of all these persons should have expressed itself in rather extravagant terms. We must doubtless discount a little the praises that are lavished on the vestals at the base of their statues; but they have the merit at least of making us acquainted with the qualities that were expected of them. They are especially praised for the zeal and skill with which they perform their sacred duties. They are said to have watched devotedly day and night at the foot of the altars of the gods beside the eternal fire, and their prayers are supposed to have contributed much to the prosperity of the republic. Many of the

---

[1] *Corp. insc. lat.*, VI. 2131.
[2] Lanciani, No. 6.

virtues for which they are lauded, such as chastity,
piety, strict observance of rules, devotion to duty, would
have applied equally well to the Christian nuns, but a
Christian would not have allowed the magniloquence
and exaggeration of some of the compliments. She
would have blushed to have it said of her that "she
surpasses all women that have gone before her in
devotion and goodness," or that "the goddess had
reserved her for herself, and had chosen her out
especially to be consecrated to her service." We can
well believe that those who lavished such praises on
the vestals made sure of not displeasing them, which
proves that humility was not one of the virtues on
which they prided themselves. We remark that one
of them is said to have been renowned for her wonderful learning (*doctrinæ mirabilis*). We know for a fact
that the worship of Vesta was a very complicated affair,
and a long initiation was necessary in order to be able to
carry out the rites according to the prescribed forms. The
thirty years for which a vestal bound herself were divided
into three equal periods: during the first she learned
her duties; during the second she performed the service;
and during the third she taught the novices. Indeed,
we see on one of the pedestals, which have been found
in the *Atrium Vestæ*, a young priestess thanking an
older one for the good lessons which she had taught her.
Another of these monuments presents a very remarkable peculiarity; the name of the vestal to whom it
has been raised has been so carefully obliterated that it
is quite illegible. If so much trouble has been taken
to efface it, it must have been because the vestal was no

longer considered worthy of the honour that had been done her; and one immediately thinks that she must have broken her vow of chastity, which fault was always punished with great severity. Another, and more plausible idea, has been put forward. The pedestal bears the date of the consulship of Jovian and Varro, that is to say, almost immediately after the death of the Emperor Julian, just when the struggle between the two religions was most violent. Should the chief vestal have abjured her vows, the affair would have made a stir, and it might have been considered indiscreet to have made the matter public. So we are led to believe that her fault was one of another nature, and, as the poet Prudentius speaks of a vestal who about this very time was converted to Christianity,[1] we may reasonably suppose that it might be this one. If the conjecture is true, the rage of the followers of Vesta, and the care they took to destroy the name of the culprit, may easily be understood.

The large court of the *Atrium Vestæ* has been cleared, and it now presents a most curious aspect. All the fragments of statues which the excavations have brought to light have been arranged along the walls in the very places where the statues of the chief vestals stood when the place was intact. Thanks to these ruins, it becomes easy to repeople this desert peristyle in imagination, and to restore to these vast porticoes their ancient inhabitants. The portraits of the vestals which remain

---

[1] Prud., *Peristeph.* II. 527: *Ædemque, Laurenti, tuam Vestalis intrat Claudia.*

to us permit us, mutilated as they are, to get an idea of what they must have been like, as well as all the details of their severe and rich attire. We recognise the short hair, bound with the *infula*, from which hung short fillets, forming a sort of diadem round the head; the cords which confined the tunic at the waist, and the round *bulla* which was worn on the breast in the same manner as that in which nuns wear the cross. M. Lanciani observes that this dress gave them quite a regal appearance, and we must confess that their dwelling was much more sumptuous than any of our modern convents. Let us not forget that the court which we are now visiting, and which must have been much frequented by them, was 68 mètres long by 20 wide. When we think that the house was only occupied by six or seven vestals, these dimensions may well surprise us; but M. Jordan accounts for them in a very ingenious way. According to him, certain indications seem to point out that a part of the peristyle was arranged like a grove, with trees, paths, and marble seats. This arrangement was not only to give pleasure to the vestals, and to make them more contented with their life, but it was really a necessity to them. "We must not forget," says M. Jordan, "that they belonged to the first families in Rome; that the class from which they came were accustomed to pass the hot months in the country among the mountains or at the seaside; they, on the contrary, having once entered the *Atrium*, found it difficult to get very far away from it again. Their duties kept them in the neighbourhood of the Temple of Vesta, and they had to say good-bye to Tibur,

Præneste, Tarentum, and Baiae. In early times this confinement was a little more endurable to them; between the *Nova Via* and the Palatine, there was a sacred wood called *lucus Vestæ*, which is mentioned by Cicero.[1] But it soon disappeared; in this part of Rome, which became every day more and more thickly populated, there was not an inch of ground that was not built upon; air and light became ever more scarce, and the poor vestals who were forced to live among this accumulation of walls endeavoured to procure at home what the neighbourhood could no longer provide for them. Thus it was that a spacious dwelling came to be made, where it was possible for them to obtain fresh air; and a little garden was laid out to delight their eyes with its fresh verdure. It was not much; but in this respect the ancients were content with little; and the masters of the world, established close to them upon the Palatine, were themselves not much better off. A grove is not worth much without a fountain; and one has been found in the *Atrium Vestæ*. It is a basin 4.40 mètres by 4.10 mètres, which even now is lined with marble. It has caused much surprise to find that, neither in the basin itself nor in the environs, has any trace of an aqueduct been found which might have brought water to the fountain when it was necessary to fill it; but M. Jordan has accounted very reasonably for this peculiarity. Festus says that the vestals did not use any water that did not come from an absolutely pure source, and they were forbidden to avail themselves of

---

[1] Cic. *De divin.*, I. 45.

that which the water-pipes brought from outside.[1] We
must suppose, then, that every morning the numerous
slaves attached to the house brought water from some
neighbouring spring and poured it into the basin. A
conduit has been found which permitted it to flow into
a drain which passed under the building.

As was always the case in Roman houses, all the
sitting- and bed-rooms were disposed round the court.
According to custom, the reception room or *tablinum*
was placed at the end, opposite the basin. This is a
very large room, which must have been richly decorated;
it is surprising that it was not placed in the centre.
This peculiarity can only be explained by the repairs
that have been made to the monument at various times
when the arrangements must have been changed. The
other rooms are in ruins, and it is difficult to say to
what purpose they were applied. It seems, however,
that the vestals must have used some of them for work-
rooms in which, for example, they made the *mola salsa*;
others were reserved for their private use. These must
have been the apartments arranged along the porticoes
on the Palatine side. Some, a little better preserved
than others, still retain their wall-facings of precious
marbles with stucco friezes which have not lost their
brilliant colours. While I was curiously examining
them, and admiring the richness of their decoration, I
could not avoid thinking of the famous quarrel between
Symmachus and St Ambrose over the altar of Victory.
Symmachus bitterly attacked the laws which the last

---

[1] Festus, pp. 158-160.

Emperors had made against the pagan priests. He pitied the vestals more than any; he spoke with emotion of "those noble maidens who have consecrated their virginity to the welfare of the State," whose property had been taken from them, and of the treatment they had received from the public Treasury. St Ambrose, in reply, insinuated that these "noble maidens" were not altogether worthy of the admiration which Symmachus expressed for them. He recalled with pleasure their privileges, their fortune, the consideration with which they had been surrounded, the large allowance which the State had made them; and hinted that there were only seven of them to share all these advantages. "All that we can call to mind about the Temple of Vesta is the honour of the fillets with which the vestals cover their heads, the splendour of their purple vestments, the litter in which they are carried, the train of servants which follow them, the immunities granted them, their liberal allowance of money; and, lastly, the right they have of not binding themselves for more than a certain number of years!" With these few great ladies, blessed with all the gifts of fortune and enjoying all the pleasures of life, he compares the Christian nuns, so simple, so humble, and, at the same time, so numerous, whom he calls by the beautiful phrase *plebem pudoris*. "They have no rich fillets, but wear an ugly veil over their faces. Instead of trying to enhance their beauty by all the tricks of dress, they affect a most simple attire. What they desire, what they seek, are not the pleasures of life; it is fasting and poverty." He was certain that this contrast between the Christian monasteries of this time and the aristocratic convent of these vestals would be a striking one. It seems to me that a visit to their sumptuous house as we know it since the last excavations have been made, and the sight of the apartments of which such beautiful fragments remain, must form a commentary to the words of St Ambrose.

Let us leave this rich and vast peristyle where we have lingered so long. A staircase of twenty-six steps brings us on a level with the road, of which we can follow the course from the Church of St Maria Liberatrice to the Arch of Titus, and which passes close by the side of the *Atrium Vestæ*. It is thought to be the *Nova Via*, mentioned more than once in Roman history, and which bordered the Palatine Gate and the Temple of Jupiter Stator. It must be confessed that the seclusion of the vestals could not have been very strict on this side, and access might have been very easily gained by means of the low windows. A few more steps lead us to other rooms, of which the mosaic pavements are the only remains. Some of them must have been bathrooms; brick pipes are still to be seen in the walls, which must have served to carry water into the marble baths. In the centre of these apartments, which appear to have been clumsily repaired in the last days of the Empire, are the fragments of another staircase, which proves that the rooms of the vestals could not have been lower than the second storey.

It is from here that we get the best view of the Palatine, as it appears since the last excavations. Those who have not visited it for two or three years will have some difficulty in recognising it. Until lately the Palatine was separated from the Forum by a dusty road leading to the entrance to the Farnese Gardens. Then, when it had passed under the gate built by Vignolius, it ascended, terrace by terrace, to the Palace of the Cæsars. This road is now a thing of the past. The mass of rubbish and earth that covered up the ancient houses has been removed, and the ruins that have been so long hidden have been brought to light. From the top to the bottom of the hill nothing is now seen but stone or brick walls of unequal height, and the framework of houses. This spectacle, I fear, will not be to the taste of everyone; more than one artist will perhaps find fault with the archæologists, and reproach them bitterly for having replaced the Farnese Gardens, from which such beautiful views were obtained over the *Campo Vacchino*, by something which resembled the streets of Paris when it was half destroyed. Certainly, archæology cares very little as a rule for beauty—it is content with truth; but truth has its charm too. Perhaps, on looking at the Palatine as the new excavations have left it, the eye is at first bewildered by the accumulation of ruins; but imagination soon sets to work. It raises vanished houses on the shapeless ruins, it joins broken walls, it erects houses that have been destroyed, and soon shows us this quarter as it must have been towards the end of the Empire.

We have more than one lesson to learn from the curious spectacle which it presents. We see once more how little the ancients cared for the wide streets and open spaces which our modern towns could not do without. We are here at the foot of the Imperial palace, a short distance from the Forum—that is to say, in the heart of the great city; and yet we have before our eyes nothing but a mass of houses creeping up the hill, jostling each other to suffocation, and leaving no empty space between them. The two parallel roads which separate them, and which run along the side of the Palatine—the *Nova Via*, of which I have just spoken, and the *Clivus Victoriæ*, a little higher up—were not enough to give the light and air of which this quarter was in such need. It was impossible to keep off the invasion of the houses which bordered them. These dwellings encroached little by little upon the footpath; then, after having almost met down below, their top stories were united by small arches thrown across the street from one roof to another, and on these arches were built aerial rooms; so that in time the *Nova Via* and *Clivus Victoriæ* became dark haunts for cut-throats. It struck me as I walked along it that it was doubtless in such a street that, in Sylla's time, Sextus Roscius was killed by the assassins as he was returning from dinner. (*Occiditur ad balneas palatinae rediens a cena.*[1])

The other observation, which the sight of this new quarter suggested to me, regards the palace of the Cæsars. Formerly, when the only entrance was through the Gate of Vignolius, when these grand ruins were separated from the Forum by fields and walls, this building gave one the idea of an isolated and closely secluded dwelling. It is the general idea that one

---

[1] Cic., *Pro. Rosc. Amer.*, VII.

always has of a king's palace. But this is not so; the new excavations show us that we have made a mistake. The house of Caligula, who was perhaps the most superstitious of the Cæsars, almost touched the other houses on the hill. Thence a staircase, still almost intact, takes one down into the centre of the *Clivus Victoriæ;* then from the *Clivus* it continues as far as the *Nova Via,* which we know met the Forum; in this way it was possible to ascend directly and in a very few minutes from the *Via Sacra* to the house of the prince. There is nothing here which resembles the dwellings of Eastern despots as Herodotus paints them for us, defended by their many enclosures and their entrenched camps. Nothing separates the houses of Augustus and Tiberius from the others; they live in the midst of the people, and are not separated from the rest of the Romans by moats and walls. This is done so as to make the people believe that they were citizens as well as themselves, to persuade people who judge by appearances — and the great majority do so — that the Cæsars must not be considered as kings, and that, under their rule, Rome was always a free city.

So we possess two out of the three monuments which recall the most ancient religion of Rome—the temple where the sacred fire burnt, and the dwelling of the vestals. The third one alone remains to be discovered, the *Regia*, that is to say, the residence of the high priest, where Julius Cæsar dwelt. Must we believe, with Signor Lanciani, that the Regia disappeared long before the ruin of the Empire ? or must we think with

M. Jordan that it will be found under the Church of St Maria Liberatrice? The future alone can say.

We have now arrived at the entrance to the Forum, whither the road debouching near the Temple of Antoninus, by whatever name it be called, leads us. Before entering and trying to describe it, I think it as well to detain the reader yet a moment on the threshold. There are a few important reflections to be made at the outset, if we would avoid serious disappointment. Let us not forget that the Forum we are about to visit is that of the Empire. Most of the monuments of the epoch of the kings or of the glorious times of the Republic, which we are tempted to seek before all the others, are no longer there. It has been so often reconstructed and altered, it has so many times changed its appearance that those ancient memories have left very little trace on it. They only exist for us in the texts of the old writers who tell us of them. But those texts, although obscure and rare, have been interpreted with so much sagacity by a learned criticism, that we are now able, without great trouble and with sufficient probability, to replace those poor monuments of Rome's earliest times upon the ground encumbered with buildings of another age.[1]

The aspect and natural configuration of the place

---

[1] For the better understanding of what follows, I have reproduced, with a few slight modifications, the map given by M. Detlessen at the end of his work on the *Comitium*, in *les Annales de l'Institut de correspondence archéologique* (1860). Although only the primitive Forum is here in question, it was not possible to make the sites of the more ancient monuments intelligible without marking those of the following epoch,

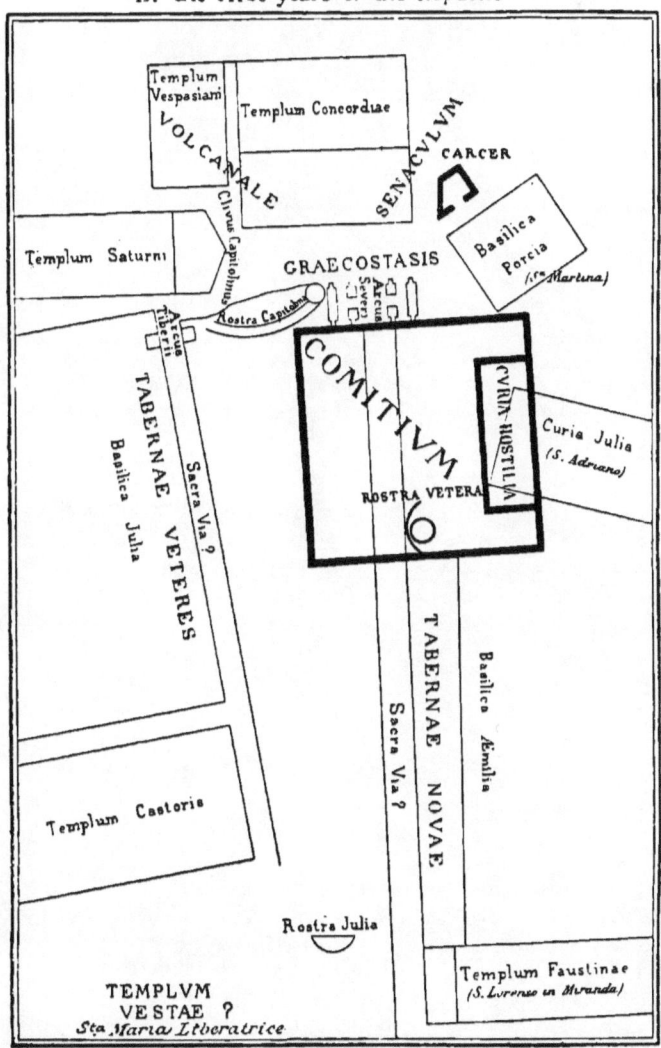

THE FORUM
in the first years of the Republic

After Detlefsen

greatly help us in this. We have seen that, according to Denys of Halicarnassus, Romulus and Tatius used to meet in a certain part of the Forum in order to confer, and that at this spot, since called the *Comitium* (gathering), the citizens thenceforth held their assemblies. But where was the site of the *Comitium* to be looked for? For a long time it was customary to locate it a little everywhere—even in the lowest parts of the plain. Good sense, however, tells us that it must have been in a high place, safe from floods. The Forum, in its primitive state, was a marsh.[1] Tarquin, by building the great drain discovered under the portico of the *Basilica Julia*, caused the stagnant waters of the Tiber to flow off, and first rendered the bottom of the place practicable. Before his time, there could have been no question of establishing a place for public meetings there. We must therefore put the *Comitium* a little higher, on the slope of a hill, in a dry spot. The texts of the old authors prove that it was to the north-west of the Forum, towards the part where we now find the Arch of Severus and the churches of Santa Martina and St Adrian. It formed a square, raised a few steps, surrounded by a balustrade, and sufficiently extensive for the curial *Comitia* to be held there. Above the *Comitium* was built the *Curia*, where the Senate met.

---

but, in order to avoid all confusion, they have been given in thinner lines and smaller letters. Of course, in an attempt to go back to such remote times, of which scarcely anything remains, minute exactitude cannot be expected. Detlessen's map only tries to give us an approximate idea of the Forum in the regal and republican epoch.

[1] Ovid, *Fast.*, VI. 401: *Hoc, ubi nunc fora-sunt, udæ tenuere paludes.*

It is unanimously believed that it was situated at about the place covered by the church of St Adrian. A little higher than the *Curia* a somewhat extensive platform was occupied by different public monuments, notably by the Græcostasis, where foreign ambassadors used to wait until the Senate should deign to receive them, and by the Temple of Concord, whose remains still exist, and serve to fix the position of all the rest.[1]

So we can picture to ourselves the ancient Roman Forum, although scarcely anything now remains of it. Let us imagine, at the foot of the Capitol and of the citadel, a series of terraces rising one above the other. Lowest of all, we have a sort of swampy plain, the real Forum, where the plebeians meet; a little higher is the *Comitium*, a square esplanade serving as a place of assembly for the nobles—that is to say, of the real citizens who govern Rome—while higher yet we find the *Curia*, where the senate holds its sittings, and of which the *Comitium* is, so to speak, the vestibule :[2] so that the very configuration of the place is an exact image of the political constitution of the country, and the various stages in which it is distributed represent different degrees of the social hierarchy, each class ascending higher as, in fact, it raises in power the nobles above the plebeians, and the Senate above all.

This State, so severely kept, where all the classes of

---

[1] Pliny, XXXIII. i. 6 : *Ædem Concordiæ . . . in Græcostasi, quæ tunc supra comitium erat.*

[2] Titus Livius, XLV. 24 : *Comitium vestibulum curiæ.*

society are so well subordinated one with the other, is not, however, a despotic State. The Aristocracy, which holds the power and desires to keep it, does not resemble that of Venice, which deliberated in the dark and forbade liberty of speech. The gravest questions are handled in the *Comitium*, in the light of day, and everything is carried on by word of mouth. In the place where public meetings are held, there is a tribune for the orators, and it is regarded as a sacred spot (*templum*). It is a small terrace of some little height and breadth, and without any balustrade, where he who speaks is completely seen from all sides, which obliges him to drape himself becomingly and assume noble attitudes. The wall supporting it bears a singular ornament: the iron prows (*rostra*) of the ships found by the Romans in the port of Antium, after the taking of the town, have been fixed there. They burnt the ships, not knowing what to do with them, and brought away the *rostra* as a trophy to decorate their Forum. The site of the tribune can be fixed with sufficient exactness. We are told it was close to the *Curia* :[1] the Senate, aware of the importance of speech, desired closely to supervise it. "It has its eye on the tribune," says Cicero, "and holds it in hand, to restrain it from rashness and keep it in bounds."[2] A passage in Pliny informs us that it must have been situated opposite the *Græcostasis*, that is to say, on the

---

[1] Asconius, *Cic.*, *Pro Mil.* 5 : *Erant enim tunc rostra non eo loco quo nunc sunt, sed ad comitium, prope juncta curiæ.*

[2] Cic., *Pro Flacco*, 24: *speculatur atque obsidet rostra vindex temeritatis et moderatrix officii curia.*

other side of the church of St Adrian.[1] Finally, we know that it was at the extreme limits of the *Comitium*. Thence the orator can be heard by everyone, and his voice reaches the different degrees of the Forum.[2] Only during the first centuries he is obliged to turn towards the *Comitium* when he speaks. He must preferably address the noble assembly which really governs the town. Later on, Licinius Grassus, or, according to other authors, the Gracchi, dared to violate this ancient usage, and first turned towards the Plebs. Sovereignty had changed place.

The Forum being the most frequented place in the town, commerce naturally flowed thither. It is said to have been surrounded by shops as early as the period of the kings. The western side, opposite to the *Comitium*, offered more free space, and was thus the first to be built upon. There arose what were called the "old shops" (*tabernæ veteres*). When ground failed on this side, they crossed over to the other, and on the space left vacant by the *Comitium* and the *Curia* erected the "new shops" (*tabernæ novæ*). Very different trades must have been carried on in them, at least in the earliest times. The school, to which Virginia was going when she was seized by the people of the triumvir Appius, was situated in the Forum. We are told that, when her father was forced to kill

---

[1] Pliny, *Nat. Hist.*, VII. 60. He says that in order to determine the hour of noon, they looked at the sun between the *Græcostasis* and the *rostra*. The *Græcostasis* being close to the Temple of Concord, to the right of the *Curia*, the *rostra* must have been placed to the left.

[2] Dion Cassius, XLIII.

her in order to save her honour, he went to the new shop and took a knife from a butcher's stall. Later on, the tradespeople, driven from the Forum by the fine buildings which were erected there, took refuge in the vicinity. A great number of them settled in the quarter of the *Via Sacra*. Beside the vendors of fruits and other eatables,[1] there must have been more elegant shops—perfumers, goldsmiths, jewellers. Here about the time of Julius Cæsar—that is to say, before Christianity—lived that "jeweller of the Sacred Way," to whom, in his epitaph, is awarded the beautiful eulogy that he was compassionate and "loved the poor."[2]

The old Forum, which had remained the same throughout five centuries, underwent a great change in 570, when Cato built the first basilica there. This preserver of ancient customs was often a revolutionist, who did not scruple to introduce new ones into the city; and the enemy of the Greeks hesitated not to imitate them when he found it useful to do so. He desired, above all things, to please the populace, whose favourite candidate he was. The people, for its pleasure or its business, frequented the Forum a great deal; but the Forum was not always a very pleasant place. It is often extremely hot in Rome, and it not unfrequently rains there. On the rainy and the hot days the busy and the idle knew not where to shelter themselves

---

[1] The fruit vendors of the *Via Sacra* were renowned. See Varro, *De re rust.*, and Ovid, *Ars. am.*, II. 265.
[2] *Corp. insc. lat.*, I. 1027.

in this uncovered spot. It was in order to give them a place of refuge that Cato built his basilica. Monuments of the kind served, as is known, for many uses. Not only was it customary to buy and sell and render justice there; but people often assembled in the basilicas without having anything to do; merely to chat, and play, and laugh together. It was natural that this people, fond as they were of amusements, should be most grateful to those who provided them with such places of meeting and rendezvous.[1] Unfortunately, this mode of pleasing them was not within the reach of all fortunes. A basilica could only be constructed after the purchase of the shops and houses of private persons, and these, situated in the finest quarter of the town, had assumed a great value. Cicero, who busied himself a great deal with the basilica which Cæsar intended to build, relates that the ground cost 60,000,000 sesterces (12,000,000 francs). "The owners," he tells Atticus, "were unmanageable."[2] But the favour of the people brought so much with it, that it could never be purchased at too high a price. And this is why the Forum was, little by little, embellished with the superb monuments whose remains have been restored to us by recent excavations.

---

[1] Cicero, *Ad. Att.*, IV. 16: *Nihil gratius illo monumento nihil glariosius.*

[2] Cicero, *Ad. Att.*, IV. 16: *Cum privatis non poterat transigi minore pecunia.* (On the basilicas of Rome, see Jordan, *Top.*, II. 216.)

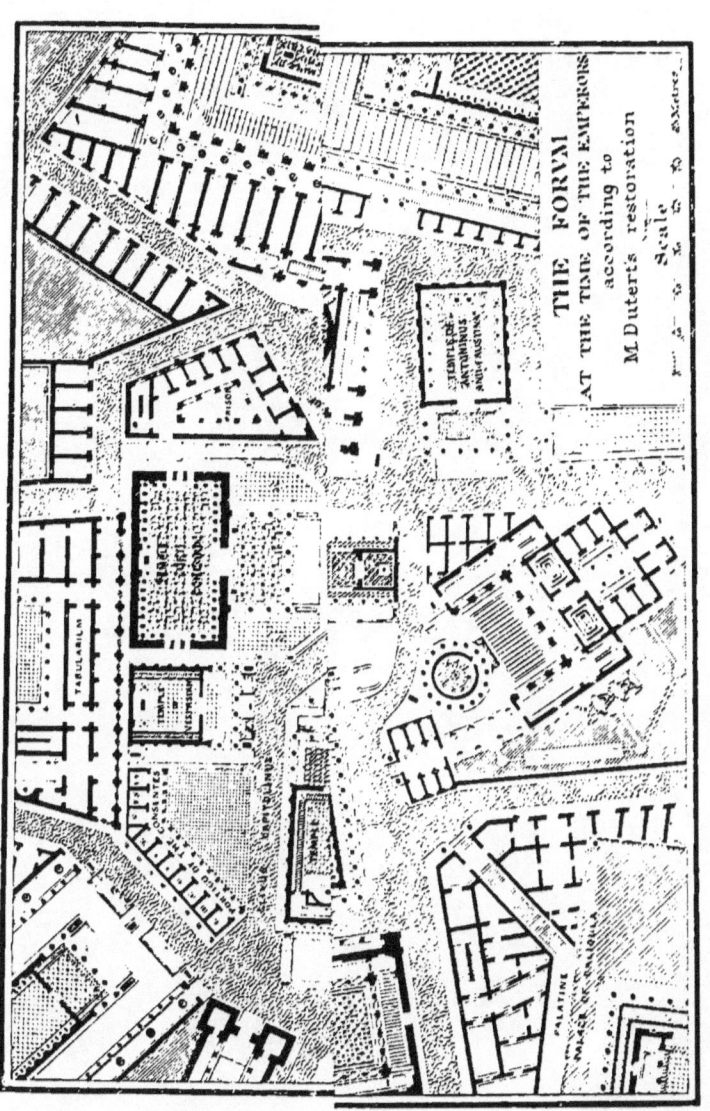

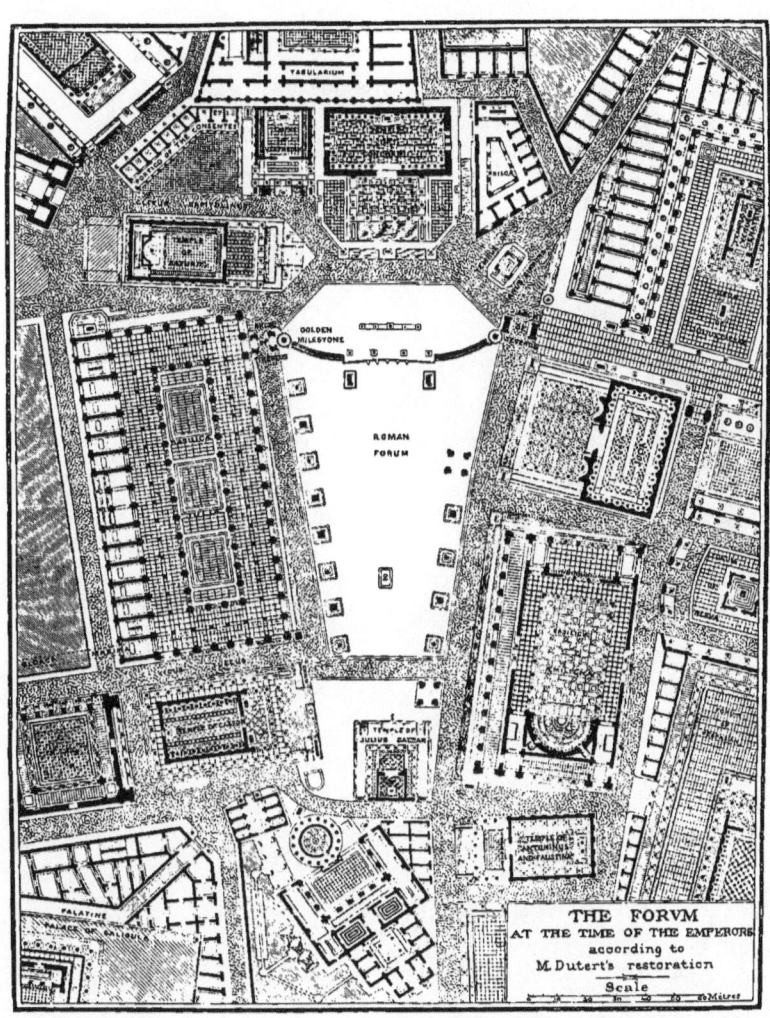

III.

THE FORUM OF THE EMPIRE—HOW WE HAVE BEEN ENABLED TO RECOGNISE AND DESIGNATE ITS CHIEF MONUMENTS—STATIUS AND THE STATUE OF DOMITIAN—THE TEMPLE OF CÆSAR—THE BASILICA JULIA—TEMPLES OF SATURN AND CASTOR—THOSE OF VESPASIAN AND CONCORD—EAST SIDE OF THE FORUM—CENTRE OF THE FORUM—THE *CLIVUS CAPITOLINUS*.

WE may now study it as it is, and, raising the ruins that cover it, picture to ourselves what it must have been at the end of the Empire. Let us go in by the newly-discovered road passing along the Temple of Romulus and that of Antoninus. At the entrance, between the latter monument and the church of Santa Maria Liberatrice, situated at the foot of the Palatine, we meet with the ruins of a building of great extent. Only the substructions remain, but they suffice to show us that it must have been a temple. The façade, which was turned towards the Capitol, presents a curious construction. The steps are not continuous, as is usually the case, the middle being occupied by a wall of peperino, rising between two narrow stairways.[1] This

---

[1] This arrangement is met with again at Pompeii. The flight of steps leading to the Temple of Jupiter, at the end of the Forum, exactly resembles that of the temple at Rome, of which we are speaking.

wall supported a sort of platform, from which a fairly complete view of the Forum is obtained. Let us place ourselves on this convenient and central spot, and thence survey the spectacle which unfolds itself before our eyes.[1]

It would not surprise me were the first glance not to fulfil our expectation. In order to join the two quarters of the modern town, it has been necessary to leave an ugly road in the midst of the excavations, called the Bridge of Consolation. This divides the Forum into two parts, and allows no portion of it to be excavated in its entirety. It is impossible to imagine it as it must have been, without first mentally removing this inconvenient obstacle. This effort accomplished, another remains to be made. We have only ruins before us, often shapeless. These heaped-up fragments are somewhat unpleasing to the eye, and in order that they may move the imagination, we must be told to what building they belonged, and know their names and histories.

After many groupings and uncertainties, this has at length been done, and the learned are now nearly agreed as regards the designation of the various monuments of the Forum. I shall content myself with giving the most important texts on which these designations are based.

In the reign of Domitian it occurred to the Senate,

---

[1] See No. 1, on the plan of the Forum according to M. Dutert. It is the spot where I suppose the observer placed to survey the Forum.

who knew their lord to possess a dainty appetite for honours, to raise a colossal statue to him, as had been done to Nero. It was placed in the midst of the Forum, and Statius, the courtier-poet, sang its erection in verses, in which, setting modesty and truth at defiance, he congratulates Domitian, above all things, on his gentleness, puts him far above Cæsar, and supposes that the old heroes of the Republic come to pay him compliments. Happily, in the midst of these repulsive platitudes, he manages to render us a signal service. In describing the statue, he enumerates the buildings round about it, telling us their names and the places they occupy; and he does this with so much precision, that he enables us to know where we are in the midst of all these ruins. But in order to profit by the indications he gives us, we must first know which way the statue faced. Statius informs us with much exactness. "Thy head," he tells the emperor, "exceeds the highest temples. Thou lookest to see if thy palace rises more glorious after the fire that consumed it, and whether the sacred fire has not ceased to burn in the solitary asylum where it must be kept alive." This means, in other terms, that it was turned towards the Temple of Vesta and the Palatine. There, now, are the monuments in whose midst it was placed. We shall see that it was difficult to be more precise and clear. " Behind thee rises the Temple of Vespasian thy father, and that of Concord ; on one side of thee thou hast the basilica of Julius, on the other that of Æmilius. Opposite, thou gazest on the monument of him who first opened the heavenly road to our princes,"—that is to say, the temple raised to Julius Cæsar after his apotheosis.[1]

So this building, situated opposite to the statue of Domitian, and which is just the one on which we have placed ourselves to look at the Forum, was the Temple of Cæsar. This monument has a curious history. Cæsar had built a new tribune of harangues, doubtless pretending as a reason that the old one was badly placed, and that it was better to put it at one of the ends of the Forum than in the middle, in order that everybody might face the orator. In reality he wished to unroot eloquence from its wonted seat, desiring that it should lose its ancient habits and accustom itself to the order of things he was about to establish. This new tribune (*rostra Julia*) was the theatre of the terrible drama enacted after the death of the great dictator. Hither it was that the body was brought on the day of the funeral, when Antony, at the right moment uncovering the body and showing its bleeding wounds, carried the crowd away by his eloquence; and it is here that the populace, maddened by grief and anger, burnt it with benches and seats taken from the neighbouring houses. On this same spot, a few days later, an altar and a column twenty feet high were raised to him, whither they came

---

[1] Statius, *Silves*, I. 22. It is believed that the stone supports on which the Colossus of Domitian rested have been found in the middle of the Forum, near the column of Phocas (see No. 2 on M. Dutert's plan). If this be true, we must suppose that when, on the emperor's death, the statue was overthrown, the pedestal was preserved, a thing hardly probable. M. Jordan is rather tempted to think that these layers of stone, which are still visible, belonged to the famous statue of Constantine.

to offer him sacrifices. When his party had triumphed, and he had been officially made a god, the altar became a temple, which was solemnly consecrated by Augustus. Only its foundations remain to us, and this platform on which we stand is perhaps all that is left of the tribune of Cæsar, from which Cicero declaimed his *Philippics*.[1]

To our left, along the road rising towards the Capitol, our attention is drawn to the ruins of a vast edifice, the finest yet discovered in the Forum. It still bears the name of Cæsar, and is the Julian basilica (*basilica Julia*). It was begun by the dictator and finished by his nephew; but scarcely was it completed when it was destroyed by a fire, and had to be recommenced. Augustus took advantage of this to remake it larger and more beautiful. There now remains of it its marble pavement, raised several steps above the surrounding streets, and extending over a surface of 4,500 mètres. As it has kept the trace of the columns and pillars that served to support the arched roof, we are able to restore its plan. The basilica was composed of a central hall, used as a court of justice. It was large enough to contain four tribunals, which dispensed justice together or separately.[2] It is here that the

---

[1] We read in a passage of the *Philippics*, VI. 5 : *Adspicite a sinistra illam equestrem statuam.* This statue, Cicero says a little further on, was before the Temple of Castor. Well, when one was on the tribune of Cæsar, the Temple of Castor was certainly to one's left, which seems to prove that this oration was delivered there.

[2] Quintilian (XII. 5, 6) relates that when these four tribunals acted separately, and the basilica was full of noise, Trachalus, who spoke before one of them, managed to make himself heard, and was applauded from the others.

most important civil causes of the time were pleaded, and where Quintilian, Pliny the Younger, and other famous advocates of that age, obtained their most brilliant successes. This great hall was surrounded by a double row of porticoes. Porticoes were then places much frequented by both sexes in order to walk and amuse themselves. Ovid strongly recommends a young man desirous "to fight his first battles" to repair thither in the heat of the day: the crowd is so numerous and so mixed, that he will easily find what he seeks.[1] Not only young people of fashion and light women in search of adventures promenaded under the porticoes of the basilicas: many men of the people also came thither, together with the idle and the unemployed, of whom there were many in that great city where the prince and the rich undertook to feed and amuse the poor. They have left their traces on the floor of the basilica. Its marble flagstones are scratched over with a multitude of circles and squares, usually crossed by straight lines dividing them into separate compartments. These were a sort of draughtboard used by the Romans for their games. The rage for gaming among the unoccupied portion of the people was incredible. It was not always obscure citizens who took part in it. Cicero, in his *Philippics*, speaks of a man of some importance who did not blush to indulge in it in the open Forum.[2] An attempt had been made, towards

---

[1] Ovid, *Ars. am.*, I. 65 *et seq.*
[2] Cicero, *Phil.*, II. 23 : *Hominem nequissimum, qui non dubitaret vel in foro alea ludere.*

the end of the Republic, to repress this mania by a law, but this law was not observed. They played throughout the Empire, and the quite fresh marks which furrow the soil of the Julian basilica show that they were still playing in Rome's last moments.[1] The basilica must have been of a considerable height. Above the first row of porticoes there was a second, accessible by means of a staircase, traces of which are still visible. From this gallery the whole place was commanded, and it was from here that Caligula threw money to the crowd, in order to have the pleasure of seeing people smother each other in endeavouring to pick it up.[2] Hence, too, one could see all that was passing in the basilica, and follow the pleadings of the advocates. Pliny relates that in a serious case, where he was pleading for a daughter disinherited by her father, who at eighty years of age had fallen in love with a designing woman, the crowd was so great that not only did it fill the hall, but the upper galleries were thronged with men and women who had come to hear him.[3]

Having become acquainted with the *basilica Julia*, the names of the surrounding monuments are easily found out. The Emperor Augustus says, in the inscription

---

[1] Some of these figures used by the players, and which are found in such great numbers on the pavement of ancient monuments, bear curious inscriptions. Here is one that has been read in the basilica Julia : *Vincis, gaudes ; perdes, plangis.* See Padre Bruzza's interesting article, entitled *Tavole lusorie del castro pretorio* (*Bull. arch. munic.* 1877).

[2] Suet., *Calig.*, 37.   [3] Pliny, *Epist.*, VI. 33.

of Ancyrae: "I have finished the basilica begun by my father, which is situated between the Temple of Castor and that of Saturn." The surroundings of the monument, therefore, are here perfectly indicated. The Temple of Saturn, where the treasure of the State was kept, rises at the foot of the slope of the Capitol. Eight columns now remain of it, of somewhat coarse execution. They were repaired at the end of the Empire, between two invasions, and this work was done with such haste and carelessness that pieces of the shafts have in some cases been replaced upside down. The other temple, near the Palatine, is that of Castor, or of the Dioscures, called by Cicero "most illustrious of monuments, the witness of all the political life of the Romans."[1] Three columns of it remain, which have ever been the study and admiration of artists. They are yet more striking now that the excavations allow of their being viewed from a lower level. Seen from the true ground of the Forum, they appear still more elegant and bold. To complete our knowledge of this side of the Forum, we should now only have to find the Temple of Vesta, with its dependent buildings—that is to say, the dwelling, or, if you will, the convent, of the Vestals and the house of the grand Pontiff, called the *Regia*. The foundations of some circular buildings have indeed been brought to light, towards the Palatine, but they did not seem to be sufficiently important for us to recognise in them the remains of the famous

---

[1] Cic., *Verr.*, V. 72.

sanctuary where the sacred fire was kept.[1] In any case, it could not be far distant, since we know from positive texts that it was near the Temple of Castor,[2] and it is still hoped that its foundations will be discovered when it is permitted to pull down the mediocre church of *Santa Maria Liberatrice*.

Opposite to us, at the end of the Forum, rises a large, modern, and very ugly wall, forming part of the municipal palace, and resting on ancient foundations. These foundations go back to the Republican period, and an inscription found there tells us they were the work of Lutatius Catullus, who finished the Capitol after the death of Sylla. They are the remains of an important monument where the state archives were kept, and called *Ærarium populi romani* (treasure of the Roman people) or *Tabularium*. It was composed of a high basement of peperino, surmounted, according to M. Dutert, by two stages of porticoes. The whole building, which must have been lower than the modern wall and allowed the Capitol to be seen, closed the Forum majestically to the north. Below, there are two temples, whose names, it will be remembered, Statius has told us. One is the Temple of Vespasian, built by his son, Domitian, quite close to that of Saturn; three columns of it remain to us. The other is the Temple of Concord, entirely destroyed.

---

[1] It is thought that one of these monuments was the tribunal of the Prætor, called *Puteal Libonis*, where justice was administered.

[2] Martial, I. 70: *Vicinum Castora canæ Transibis Vestae, virgineamque domum.*

It was a magnificent monument, which had been turned into a kind of museum. Masterpieces of the Greek artists, graven stones, and natural curiosities were to be found there. The custom of consecrating precious objects of gold and silver to Concord, in order to propitiate her, continued under the Empire. Some of these offerings were made in favour of the emperors by devoted subjects. An inscription has been found among the ruins of the temple, praying the goddess to prolong the days of Tiberius, in which he is called the best and most just of princes.[1]

So we are now acquainted with three sides of the Forum: that towards the east alone has not been cleared. It is covered by a quarter of New Rome, and in order to bring it to light again, it would be necessary to destroy all the houses from *S. Lorenzo in miranda* (the Temple of Antoninus) to Santa Martina. It will doubtless soon be done, for the Municipal Council of Rome understands that it cannot leave its work imperfect. Happily, we know, approximately, what there is there. The texts of the ancient authors inform us of this with sufficient clearness, and a very curious discovery has nearly put it before our eyes. In the course of the excavations, made near the Column of Phocas, there were found, among works of the Middle Ages, two bas-reliefs, probably dating from the first century. Many disputes have arisen as to the subject they represent, but all admit the scene to be laid in the Forum, and that the artist meant

---

[1] Jordan, *Sylloge insc. Fori*, No. 13 (dans *l'Ephemeris epigraphica*, III. p. 227).

to represent its chief monuments. In one of them we easily recognise the Temples of Castor and of Saturn, and the *basilica Julia*, that is to say, the buildings on the western side. As the other was meant to be placed in front as its pendant, it must contain those bordering the Forum on the opposite side, the only one not yet laid bare. We recognise in it the Æmilian basilica and the *Curia* of Cæsar. Thus we now possess the elements needed for a knowledge of the whole Forum.

It is not, however, exactly the Forum which we have thus far been endeavouring to describe, but only the sumptuous buildings that surround it. The ancients did not confound these with the Forum itself.[1] They reserve the name for the interior space extending between these temples and basilicas. This place, of which, so long as it was covered with rubbish, it was difficult to form an idea, is now known to us. A part of it has been restored to us by the excavations,

---

[1] We have, then, in all, three tribunes of harangues : First, that of the Republic (*rostra vetera*), near the church of St Adrian. It seems, indeed, to have still existed under the Empire, for Suetonius says that Tiberius pronounced the funeral oration of Augustus before the temple of Julius Cæsar, and Drusus on the old tribune (*bifariam laudatus est : pro æde Julii a Tiberio, et pro rostrio veteribus a Druso*, Tiberii filio. —Suet., *Aug.*, 100). The old tribune must have still retained the *rostra* of Antium, since Augustus caused the new one to be ornamented with the spurs taken at the battle of Actium. Secondly, the tribune of Cæsar (*rostra Julia*), before the temple consecrated to him. Thirdly, finally, the one placed over against the Arch of Severus, and which some *savants* call the *rostra capitolina*. On what occasion and at what period this last tribune was constructed is not known, but Canina believed that he has found a reproduction of it in the bas-reliefs placed on the Arch of Constantine, which are of the time of Trajan. These are certainly the three tribunes of harangues (*tre*

and we can imagine the rest. It was quite shut in on every side, and surrounded by the streets on which the buildings of which I have just been speaking, looked. It is not quite an oblong, as was thought, but rather a species of trapezium, wider towards the Capitol than at the other extremity. On the slabs of peperino that cover it, rise large blocks of stone, the supports of the statues and columns by which we know the Forum to have been encumbered. Eight of them are counted along the road in front of the *basilica Julia*. Towards the top, by the Arch of Severus, a wall is remarked, still covered with a few marble slabs, and pierced with deep holes, of which a part is unfortunately buried in the mass of the Bridge of Consolation. This wall seems to have supported the platform of another tribune of harangues, and it is thought that the holes were made for the insertion, according to custom, of the spars of ships. Near this tribune, in front of the Temple of Saturn,[1] Augustus erected the "golden milestone," the centre of the Roman world, where all the highways of the Empire took their beginning.[2] At this spot all the roads of the Forum joined, to proceed together to the heights of the Capitol, and here it is, as I said farther back, that we are sure to

---

rostra) which used to be shown to strangers at the end of the Empire, and which are noted in the *Curiosum urbis Romæ*. They are found in Detlefsen's map of the Forum of the Republican epoch, which we have reproduced further back. Let us add that orators frequently spoke elsewhere than at the tribune, and that demagogues have more than once stirred up the populace from the steps of the Temple of Castor (Dion., XXXXIII. 6).

[1] Tacitus, *Hist.*, I. 2 : *Milliarium aureum sub ædem Saturni.*
[2] See No. 3 on the plan.

find the *Via Sacra* again, by whatever way it passed.
At the moment when the hero of the Triumph was
about to enter upon this steep incline, known as the
Clivus Capitolinus, a sinister train divided from the
joyous troop following his chariot. It was the vanquished who had been led throughout the day behind
the conqueror, exposed all along the streets of Rome to
the insulting curiosity of the crowd. The festival over,
they took him to the Mamertine prison to put him to
death.[1] This is the fate suffered by the two noblest
foes of Rome—Jugurtha and Vercingetorix, guilty of
having bravely defended the independence of their
country. Meanwhile, the victor, continuing his course,
passed near the little terrace where we find the elegant
portico of the "Dii Consentes," and what are believed
to be the offices of the notaries. Thence he reached
the famous Temple of Jupiter, situated near the
Tarpeian rock, and of which the foundations have
recently been discovered under the Palazzo Caffarelli.

IV.

IMPRESSION FIRST PRODUCED BY THE FORUM—ABSENCE
OF SYMMETRY—ITS SMALL EXTENT—THE VERY
DIFFERENT USES IT SERVED—POLITICAL ASSEMBLIES
—HOW ORATORS MADE THEMSELVES HEARD IN IT—
HOW IT HELD ALL THE PEOPLE WHO CAME TOGETHER
THERE.

BEING now approximately acquainted with the sites
and the histories of the chief edifices of the Forum, it is

---

[1] Cicero, *Verr.*, V. 30 : *Cum de Foro in Capitolium currum flectere incipiunt, illos duci in carcerem jubent.*

easy for us, in imagination, to repair and restore all these ruins, and picture to ourselves what the place must have been like ere time had reduced it to the state in which we see it now. Let us seek to realise the impression it would make upon us, could we behold it as it was in the last days of the Empire, on the eve of the invasion of the barbarians, when it was still the admiration of visitors.

I think that in order to be duly struck by it, we should make a slight concession, and begin, which is always a difficult matter, by forgetting for a moment our habits and our prejudices. We are wont to put first among the merits of a public place its symmetry, its regularity, and its extent. It must be owned that the Forum appears somewhat to lack these qualities. It has the defect of all things not constructed according to a fixed plan. No architect regulated its proportions in advance, and distributed the monuments round about it; it may be said to have been formed by the centuries. We have seen that it originally consisted of different and unequal grades. Overlooking a swampy plain rose the *Comitium*, which had above it the *Curia*, and then the *Vulcanal*, whence one ascended by a steep slope to the Capitol. In course of time these differences of level were partially masked by the construction of large buildings, but these edifices, erected at haphazard, at very different epochs, are not always in keeping; they are heaped together without much order, and crowded one against the other. The great personages who governed the Republic being all anxious to leave memorials of themselves in Rome's most famous spot, no space about it has remained empty. We find several basilicas there, seven or eight temples, a palace for the Senate, three passages or *janus* for men of business, and five triumphal arches. Even the part lying between these edifices, and which should have been left vacant for the use of the public, was encumbered with trophies, shrines, columns, and, above all, with statues, which, to use Chateaubriand's expression, formed a whole dead nation amid a living one. Vanity had so multiplied them, that the Senate was occasionally obliged to have some removed.[1] Among the columns were some that must have taken up considerable room. They were surrounded by a balcony commanding all the Forum, and on days when a fortunate and grateful candidate gave the people a spectacle, the descendants of those in whose honour the columns had been raised had the right to come with their families to these species of tribunes, and watch from them the gladiators and the athletes. It is to be feared, then, that at first sight the Forum may strike us unfavourably; that this accumulation of riches may weary the mind; and that we may regret not to find a little more order, symmetry, and simplicity there.

But such a first impression will hardly last, if we muse on the events and personages recalled by those edifices. There, in truth, may it be said with Cicero: "On whatever spot we tread, we awake a memory."[2] The Forum is not one of those public places found in all towns, and it would be unjust to apply ordinary rules

---

[1] Pliny, XXXIV. 6, 14.
[2] Cicero, *De Fin.*, V. 2: *Quacumaque ingredimur, in aliquam historiam ponimus.* Cicero is here speaking of Athens.

to it. We must not require it exactly to resemble others in its general plan and its dimensions, since it possesses the peculiar character and the special beauty of comprising in itself the entire history of a country. The vast number of its monuments, which at first somewhat surprised us, explains and justifies itself by that of the glorious deeds whose memory they preserve. This first æsthetic defect removed, I think that our eye will soon become accustomed to the somewhat confused spectacle, and that we shall even find in it a certain picturesqueness not met with in the solemn and cold regularity of our own great public places.

It is rather more difficult to clear the Forum from another fault with which it has been reproached; and it would seem, indeed, not without reason. What strikes one at first, on viewing it as a whole, is that it does not appear very large. On observing its slight depth and extent, one asks oneself how it could have sufficed for all the uses it served. Ancient authors tell us that it was the most frequented spot in Rome. Idlers, of whom there are always so many in large towns, made it their meeting-place. Horace relates that he was accustomed to walk there every evening.[1] He was strolling, as was his wont, along the Sacred Way, the day when he met that bore who dogged his steps, in spite of his protests, and insisted on being presented to Mæcenas.[2] Curiosity there found plenty to satisfy it. Not to speak of the quacks of all kinds, of whom there was no dearth, there were sometimes

---

[1] Horace, *Sat.*, I. 6, 133.
[2] *Sat.*, I. 9, 1 : *Ibam forte via Sacra sicut meus est mos.*

genuine exhibitions of paintings. After the defeat of Greece, the masterpieces of her Art were often exposed beneath the porticoes, or in the temples, and amateurs crowded thither to see them. Occasionally victorious generals, as a device to heighten the effect of their victories, had the battles in which they had taken part painted by skilful artists, and exhibited them in the Forum. One of these, the Prætor Mancinus, carried complaisance so far as to stand beside the picture representing his great deeds, to give explanations to those who should need them. This politeness charmed the people, who the following year named him consul.[1] At the foot of the tribune, newsmongers and politicians met. They formed animated groups, eagerly discussing the latest event; they spread alarming news; they framed laws and plans of campaign, and they spared neither statesmen who had not the good fortune to be popular, nor generals who did not snatch victory at the first blow.[2] Near the same spot, below the first sundial erected in Rome, young men of fashion used to assemble, some carefully clean shaven, others with welltrimmed beards (*aut imberbes, aut bene barbati*).[3] Not far off, near the Æmilian basilica, the exchange was held. The bankers had their offices along certain vaulted passages, called *janus*, where they were seen behind their tables, entering in their account-books the money people came to entrust to them, or that which they consented to lend on good security and at enor-

---

[1] Pliny, XXXV. 47.
[2] They called them *subrostrani.*—Cicero, *Ep. fam.*, VIII. 1.
[3] Cicero, *Pro Quint.*, 18: *Non ad solarium, non in campo, non in convisiis versatus est.*

mous interest. There, stewards of great houses, knights engaged in the farming of the public revenues, merchants, usurers, and borrowers met. Important business was transacted in the place—one got rich there quickly enough, but one got poor again more swiftly yet. How many fortunes, thought to be solid, came, as Horace says, and were shipwrecked between the two *janus!*[1]

The Forum was also occasionally used for popular spectacles, especially combats of gladiators. I need not say that it was very much crowded on those days. We learn from Cicero that this was the game preferred by the populace above all others, and to which it thronged with the greatest eagerness. In order to see it to greater advantage, people crowded not only the neighbourhood of the arena, but the steps of the temples, the terraces of the basilicas, and the streets rising to the Capitol and the Quirinal. The festival often lasted several days, and usually ended with a great repast, at which all present were regaled. Tables were raised on the open Forum, and whoever chose came and sat down to them.[2] In order that the people might see and eat more at their ease in spite of the solar heat, Cæsar had the entire Forum covered with immense awnings (*velaria*), which sheltered everybody during the two or three days the festival continued.[3] Dion tells us they were made of silk.[4] This magnificence soon became customary, and once, under Augustus, when the season was

---

[1] *Sat.* II. 3, 18. [2] Pliny, XIX. 1, 6.
[3] Titus Livius, XXXIX. 46. [4] Dion, LIII. 31.

very hot, the awnings remained spread throughout the summer.[1] A spectacle yet more common than the gladiatorial combats was afforded the curious by the funerals of great personages. The procession always crossed the Forum. Players on the flute, the trumpet, or the clarion, were seen to pass, deafening all assembled; female "weepers," tearing their faces and plucking forth their hair; the crowd of friends, clients, and servants which was always attached to great houses; and finally, those cars, or litters, bearing the ancestral images, whose number was necessarily considerable when the family was ancient. At the funeral of Marcellus there were more than six hundred. What is somewhat difficult to understand, and must have made the incumbrance incredible, is the circumstance that these funerals did not turn aside from the Forum even when it was occupied by other assemblies. This is known from a celebrated anecdote related by Cicero, which many others repeated after him. The orator Crassus was one day defending a friend of his against M. Brutus, a very bad man who dishonoured a great name, and who, after squandering his fortune, earned his living by following the trade of accuser. The affair was lively, for Brutus did not lack ability, and the ardour of his hates sometimes made him eloquent. Just on that day he had spoken with great cleverness, and loaded his opponent with the most biting railleries. All at once, while Crassus was replying, the Forum was crossed by a funeral train. It was a lady of the Brutus blood being borne to the

---

[1] Dion., LIX. 23.

pyre, surrounded by all the images of her ancestors. Crassus, quick to seize the opportunity, turned to his rival: "What dost thou calmly seated there?" he asked; "what tidings shall that aged woman give of thee unto thy father, to all those great ones whose portraits thou beholdest, and to that Lucius Brutus who freed the people from the yoke of kings? In what work, what glory, what virtue, shall she say thou art busied?"[1] And he went on to reproach the unworthy descendant of so great a family with all his conduct, and with all his life. Thus did a great Roman orator find among the sights offered by the Forum to its frequenters the occasion for one of his finest rhetorical efforts.

But what above all drew the crowd to the Forum were the political assemblies. Those which met there were of three kinds: Firstly, the legislative *comitia*, where laws were voted; secondly, the ordinary meetings (*conciones*), where there was nothing to vote, and which were convened by a magistrate who had some communication to make to the people; thirdly, political suits, pleaded in the presence of everybody, before a jury drawn by lot and presided over by the

---

[1] *De orat.* For this passage of Cicero, I have borrowed the translation of M. Villemain. He has introduced the anecdote in his *Tableau de la littérature du dix huitième siècle* in a style slightly imaginative perhaps, but very interesting. His narration, which produced a great effect, commences thus: "*Voyez d'ici le Forum tel qu'il ne l'est plus, cette place immense, arène journalière du peuple roi,*" etc. There is more imagination than truth here, and we have just seen how far the Forum is from being "an immense place," etc. What M. Villemain describes is not "the Forum as it is no more," but "the Forum such as it never was."

prætor. Of these three kinds of meetings, the first—that is to say the legislative *comitia*—were the most important, and they were also the most rare. However great the mania of free people to continually alter their legal systems, there cannot be laws to make or unmake every day.¹ I add that it was not perhaps the one to which people repaired with the greatest eagerness. Those great serious speeches, in which general ideas are developed and interests of State discussed, are less suited to popular gatherings than to limited meetings, comprising only the enlightened. The multitude usually takes very little pleasure in them; they are too calm and too cold for it. In Rome, in order to arouse it, a personal question must be mixed up with the debates; and hence the importance given there to political suits. They were as frequent as at Athens, and men of state passed their lives in accusing each other and defending themselves. Parties had no other means of attack than reciprocally to bring their chiefs to justice. The scenes in which a great personage, surrounded by his weeping family, his clients, and his friends, came upon the Forum to defend his honour and his fortune were most dramatic, and so the crowd was very anxious to assist at them. It was not less numerous at the assemblies convoked by the magistrates for the purpose of communing with the people. The democracy is everywhere very exigent and very suspicious; and in Rome, as elsewhere, it required that all whom it had nominated to public charges should render it a strict account of their

---

¹ Of all Cicero's orations that have been preserved to us, only a very small number—two or three—were pronounced before the people in order to counsel it to vote a law or dissuade it from doing so.

conduct. It was a duty in which they did not fail when they desired to retain its confidence. Cato, one of the most accomplished types of the popular magistrate, always kept himself in touch with his constituents. He assembled them continually to relate to them in detail what he had done; above all things, he told them his opinion with that droll animation so pleasing to the people, and talked to them of others and of himself, without regard for his adversaries, whom he loved to call profligates and rogues, while he never tired of praising his own sobriety and disinterestedness. The people took great pleasure in these communications, which made it feel its sovereignty. In moments of public excitement, when it was known that a tribune was to speak against the Senate or handle some burning question, artisans deserted their work, shops were closed, and from all the populous quarters they crowded down to the Forum. On these days the Forum, encumbered with people, must have seemed very contracted. This was still more the case when those legislative *comitia* assembled of which I have just spoken. It was then necessary to take certain precautions with regard to the vote, to divide the place into thirty-five separate compartments in which to enclose the tribes; and to construct those narrow passages called bridges, where the citizens could only pass one at a time to deposit their voting-tickets in the baskets. When we cast our eyes over the Forum as it exists to-day, and see the small space it occupies, it is indeed difficult to understand how it should have sufficed for all those complicated arrangements and contained the assembled Roman people.

THE FORUM.

It is true, as we have already said, that this Forum which we have before our eyes is not quite the Forum of the Republic, but that of the Empire. It is sometimes supposed that its size was only diminished under the Empire, and it is added that it might then be so without inconvenience, the people no longer having laws to vote; but this supposition is not exact. With the exception of the columns and statues which continued to encumber the centre of the place more and more, and of the triumphal arches which narrowed the adjacent streets, the new buildings were constructed on ground belonging to private people, outside the limits of the real Forum. Far from diminishing its extent, Cicero distinctly says that they enlarged it.[1] On days when the assemblage was numerous, the people could group themselves on the steps or in the vestibules of the temples.[2] Those who had not been able to find room near the tribune, packed themselves together in the two stories of the basilicas. Hence they could see very well, and, strictly speaking, they could hear. It is therefore a mistake to think that the monuments built about the Forum ever prevented popular assemblies from being held there, and that it contained more people before they were constructed.

There is, moreover, a reason why it could never

---

[1] Cic., *Ad Att.*, IV. 16: *Ut forum laxaremus.*

[2] A very large number of citizens could place themselves on the steps of these temples, when, as in the case of the Temple of Castor, they were much raised above the ground. People of my age remember that in 1849, at one of the festivals celebrated by the Republic, all the pupils of the colleges of Paris—that is to say, more than 5000 children—were placed on the steps of the Madeleine, and people were much surprised at the little room they seemed to take up there.

have been so vast as our imagination loves to depict it.
It was necessary that the orators should be able to
make themselves heard in it. Whatever strength of
lung we may suppose a Cicero or a Demosthenes to
possess, it is impossible to picture them to ourselves
pronouncing their orations on the *Place de la Concorde*.
Ancient Republics found themselves in a great dilemma
when constructing their public places. They had to
make these large enough to contain a whole people, and,
at the same time, small enough for the orator's voice
not to be lost in them. Since the Roman Forum was
for many centuries the usual place of public assemblies,
we must believe that it fulfilled both these conditions.
It is a fact, and must be accepted, even though not
very easy to understand. We must first admit, then, that
orators could be heard there, even when not very well
listened to, and that their voices dominated those noisy
assemblies, sometimes compared to the waves of a
troubled sea, and where people abused each other, spat
in each other's faces, and flung stones and stools at each
other's heads. It may well be thought that they did
not attain to this without effort. They had to learn a
particular mode of emitting the voice ; to sing their orations as it were, and, above all, to accompany them with
an expressive pantomime which made them more easy
to follow, hence the importance of rhythm and gesture
in ancient eloquence. It is, thanks to these means, that
they succeeded in making themselves heard. Perhaps,
too, the situation of the Forum will help us to understand what at first appears to us a veritable prodigy.
It is placed in a kind of hollow reached by steep slopes.
Towards the Capitol, it is a genuine precipice. At the

opposite end, towards the Arch of Titus, the incline is more gentle, but still very decided, while on all sides, as it was customary to say, one "descended" to the Forum. When we think that this configuration of the locality, the small extent of the place, the hills surrounding it, the buildings that shut it in, are very favourable for the voice, it becomes somewhat less astonishing that orators should have made themselves heard in it and produced those great effects which history has handed down to us.

We must also admit, in spite of the surprise we feel, that this Forum, which appears to us so confined, might have contained all those who wished to be present at some important suit, or who came to exercise their suffrages on a voting day. Perhaps, after all, the number of these voters may have been less than we are tempted to think; possibly the place only sufficed, because a portion of those who had the right to come there remained at home. Towards the end of the Republic, in proportion as the popular assemblies became more stormy, wise and moderate people, who in all countries are the most timid, got into the habit of shunning them. When it was seen that they usually ended in sanguinary brawls, persons who feared noise ceased to appear at them. Cicero complains bitterly of this desertion of the *comitia*, and speaks of certain laws passed by just a few citizens, who had not even the right to vote. This was what explains how so many Romans so easily accepted the Empire. They cared little enough about being deprived of political rights which they had themselves renounced.

Yet under the Empire, the Forum ended by appearing too small. The popular assemblies then no longer existed, but the promenaders, the idlers, and the curious became more and more numerous, and strangers arrived from all the corners of the world. The expedient was resorted to, not of enlarging the old Forum, which could not have been done without destroying ancient monuments, but of building others round about it. Cæsar began, other princes imitated him, and, as each desired to eclipse his predecessors, the cost became each time more considerable and the constructions more handsome. Thus it was that they came to create, in the heart of this sovereign city, the finest assemblage of monuments and public places with which a town was ever honoured. A foreigner entering Rome by the Flaminian Way, and who, after passing through the Forums of Trajan, of Nerva, of Vespasian, of Augustus, and of Cæsar, at length came to the ancient Roman Forum, where the beauty of the edifices was enhanced by the greatness of its memories, must have been strangely surprised at this sight. However great an idea he might have formed for himself in his own country of the marvels of Rome, he was obliged to own his dreams far below the reality. He well felt that he was in the world's capital, and he went home full of an admiration that faded not for the town on which the whole universe had its eyes, and which, from the second century downward, was only spoken of as the "Sacred City."

CHAPTER II.

THE PALATINE.

The excavations of the Palatine, like those carried out on the Forum, have led to very curious discoveries. This hill, formerly occupied by the villas of great lords and the gardens of monasteries, where nobody might penetrate, has become one of the most interesting walks of Rome. I do not believe there is a spot where recollections of the past so crowd upon the memory, and where one more lives in mid-Antiquity. It must, however, be owned that this Antiquity was only given back to us in a very sorry plight, and persons allowing themselves to be beguiled by the tablet placed above the entrance to the Farnese Gardens, and believing that they were really going to find the "Palace of the Cæsars" again, would run the risk of being greatly surprised on seeing what really remains of it. There are only a few ruins left, and, in order to see it such as it was, we must make a great effort of imagination.

This effort is, however, necessary almost everywhere in Rome if one would feel some interest in visiting it. Everybody about to journey thither should be told this, in order to spare disappointment. Rome is not quite like other Italian towns—Venice, Naples, or Florence, for example—which impress the visitor at once. It does not produce all its effect so quickly, and, in order to fully enjoy it, a sort of initiation is indispensable. There are many reasons which prevent the great monuments it contains from at first corresponding to the idea we had of them. We hurry, on our first arrival, to see the ruins of which we have heard so much, but they are usually fixed into modern houses, and for the first moment this common surrounding prevents us from seizing all their beauty. We hasten to visit the old churches dating back to the first centuries of Christianity, but having been very often repaired and rejuvenated, they have lost much of their true character and primitive originality. One is not much struck by them on a cursory view, nor is this rapid glance enough to enable us to appreciate them as they deserve. It may be said that each year thousands of hasty travellers pass through Rome who, not giving themselves time to see it, carry away only an incomplete impression. Some, the most courageous and most sincere, dare to own their disenchantment; others admire on trust and from pre-intention, in order to do like everybody else, and thus not have lost their journey. Let us not follow their example; let us be at the pains to see again, more than once, those fine ruins which at first left us indifferent; let our imagination help our eyes to understand them; let us, in thought, endeavour to isolate them from their dull, uninteresting surroundings; let us encircle them with the great memories by which they are ennobled, and assuredly all will then change its aspect for us.

To understand and know Rome is a study, then; a

study requiring time and demanding some efforts, but this time is well employed, and these efforts promise us one of the greatest pleasures an intelligent man can give himself. Far from this pleasure being lessened by its postponement, we, on the contrary, find in it a special charm, because it is, so to speak, our work; because we partly owe it to ourselves; and because we are pleased with ourselves for what we did in order to win it. What completes it and lends it its finishing zest, is that it is coupled with a secret self-satisfaction and a certain feeling of pride that is liveliest in the most cultivated minds; that it calls for familiarity with the past and a full understanding of it; and, finally, that dunces and fools can never more than imperfectly enjoy it. Other towns, even those we most love, only make us pleased with *them;* Rome possesses the unique privilege of at the same time delighting us with her and with ourselves. Let us add that, if the pleasure felt in visiting her does not come at once, it always grows with time. In studying all these monuments more nearly, we continually find new reasons for being struck with them. The more we view them, the greater charm we find in their sight, and we end by feeling the greatest difficulty in parting with them. Rome is the town of the world where curiosity and admiration least weary, and it has been remarked that those who have lived there longest are both the least inclined to leave it and the most desirous to return. Pope Gregory XVI., who was a clever man, always asked foreigners who came to take leave of him, how long they had stayed in Rome. When they had only remained a few weeks, he merely said: "*Addio!*"

but when they had sojourned there several months, he always said: *"A rivederci !"*

These reflections, which apply to Rome generally, are, perhaps, still more appropriate with reference to the ruins of the Palatine than to all the others. It is there, especially, that the too hurried traveller is in danger of understanding nothing, and it is there that the curious lover of Antiquity, who takes time to learn, is sure to be largely repaid for his trouble. The Palatine, being the most ancient of the quarters of Rome, architectures of different epochs were more crowded together there than elsewhere. It had possessed great importance under each form of government, and the Kings, the Republic, and the Empire all left important monuments there, which for ten centuries were covered over with earth. Recent excavations have given them back to us, but, unfortunately, they have given them back all mixed up together. These edifices, having sunk one upon the other, re-appear simultaneously, and at first it appears as if one would never come to a clear understanding of them. But fortunately at Rome each century had its particular manner of building, and each epoch used different materials. According as a wall is formed of peperino, of travertine, or of brick, or the work executed in what is called *opus incertum* or *opus recticulatum*, one can approximatively tell its age. Moreover, in the manner of joining the bricks together, or laying the blocks, there are indications which do not deceive a practised archæologist. Lastly, one sometimes happens to find inscriptions on the leaden water pipes, and the bricks occasionally bear the mark of the factory whence they came, or even the name of the consul under whom they were made, thus ending all doubts. It is thus we have been enabled to distinguish with great probability the age of the monuments discovered. Let us profit by these guides in order to arrive at an understanding of what remains of the Palace of the Cæsars, and endeavour to ascertain what recent excavations have restored to us of the various historical periods of the old Palatine.[1]

I.

HOW THE EXCAVATIONS ON THE PALATINE CAME TO BE UNDERTAKEN—*ROMA QUADRATA* AND THE WALLS OF ROMULUS—THE TEMPLE OF JUPITER STATOR—REMAINS OF THE EPOCHS OF THE KINGS—ANTIQUITY OF WRITING AMONG THE ROMANS, AND THE CONSEQUENCES TO BE DRAWN FROM IT—THE PALATINE UNDER THE REPUBLIC — WHY EXCAVATIONS ARE ALWAYS SO PROLIFIC IN ROME.

THE Palatine is a hill nearly 1800 mètres in circumference and 35 mètres in height, placed like a kind of island in the centre of those which, together with it, formed the Eternal City. Although the smallest of all, "the others," says a writer, "appear to surround it with their homage, as their sovereign."[2] And it was

---

[1] We are about to enumerate the chief monuments of the Palatine according to their age, and not in the order in which they present themselves to the traveller.

[2] *A cui, come a sovrana, fan le altre sei corona.*—Guattani, *Mon ined.*, January, 1785.

indeed this hill which held the greatest place in the existence of Rome. As it was natural to believe that it must contain fine mementoes of its glorious past, it has since the Renaissance been several times excavated. According to the custom of the day, mosaics, statues, or other objects of art were sought, and the curiosity or the cupidity of the explorers once satisfied, the ruins for a moment brought to light were hurriedly covered up with earth again. Serious and continuous works were only begun in our time through the initiative of France. In 1861 it occurred to the Emperor, Napoleon III.— whose passion for Roman history, and, above all, for the history of the Cæsars, is well known—to buy of Francis II., King of Naples, the Farnese Gardens, occupying the north of the Palatine.[1] This project met with many obstacles on the part of the Roman court, which did not care to see France become so near a neighbour. It raised a thousand difficulties, which were not to be overcome without much trouble. The great French epigraphist, M. Léon Renier, who knew the importance of the acquisition, and had counselled it, had the honour to conclude the negotiations. When they were finished, and the Palatine belonged to France, he made known to the Emperor the architect who appeared to him best fitted to undertake the great works which it

---

[1] If a guide to the Palace of the Cæsars is needed, the one published by MM. C. L. Visconti and Lanciani, of which they have given a French edition (*Guide du Palatin*, Rome, Bocca), should be chosen. It is an excellent work, very clear, very learned, and very complete. As will be seen in the following work, I have used it a great deal. I have reproduced the map placed by MM. Visconti and Lanciani at the head of their book, almost exactly.

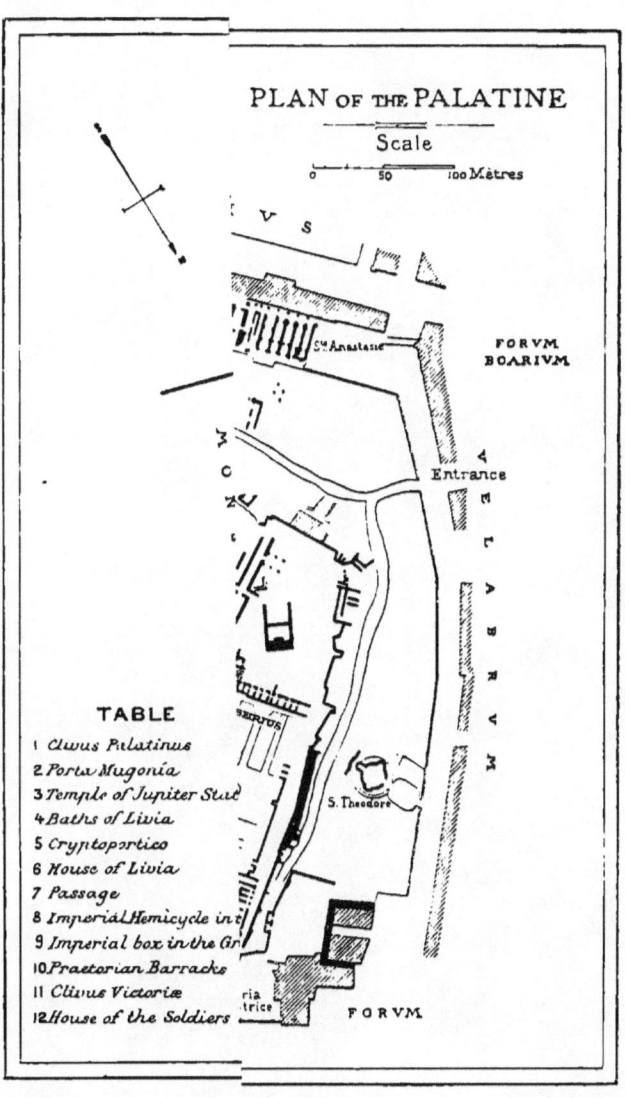

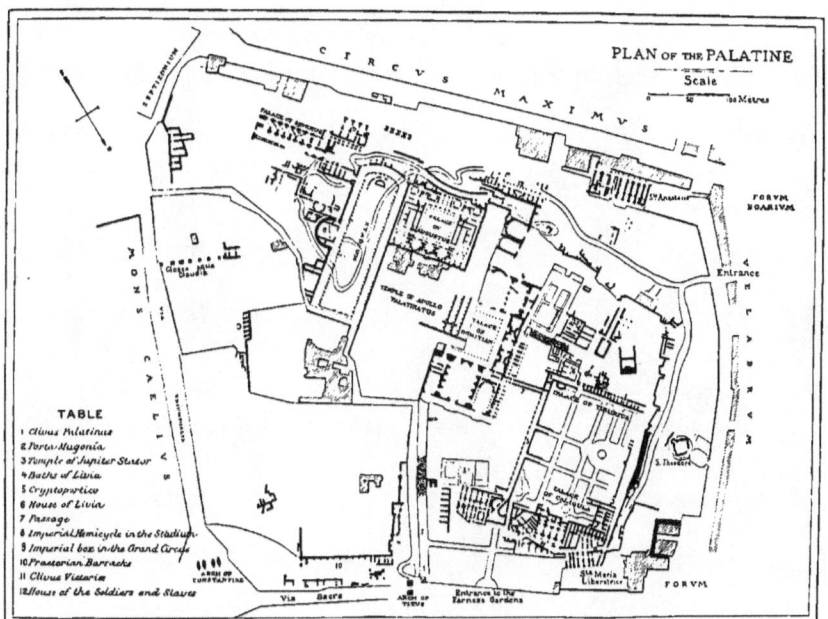

was proposed to carry out there. This was M. Pietro
Rosa, known to the learned by his topographical studies
on the Roman Campagna. M. Rosa at once set to work
with ardour, and was not long in justifying, by the most
important discoveries, the confidence shown in him.[1]

These discoveries were not limited to the Imperial
epoch. While searching, especially for the Palace of
the Cæsars, they found remains of the old town of
Romulus, which might have been thought for ever
lost. It was well known to have been built on the
Palatine. History relates how the first king, having
called around him all the adventurers of the neighbourhood, marked out its boundary in accordance with
the Etruscan rites. They say that he harnessed an ox
and a cow to the plough, and guided it all round the
hill, raising the share at the spots where the gates
were to be, and marking by a deep furrow the circumference of the town which he desired to found. This
furrow, or rather the space left free beyond the line
traced, formed what was called the *pomœrium* (*pone
muros*) or sacred enclosure of the town, within which
it was forbidden to bury the dead or to introduce
strange gods. Its limits were marked by stones placed
from distance to distance along the Palatine. In the
time of Tacitus it was believed to be still known,
and its position was pointed out.[2] This was "Square
Rome" (*Roma quadrata*), so called from the shape
of the hill on which it was seated, or rather because

---

[1] Directly after the events of 1870 Italy bought back the Palatine of
the Emperor Napoleon III., while he was still a prisoner in Germany.
[2] Tacitus, *Ann.*, XII. 24.

it had been founded according to the rules of art of the Augurs, and represented on the earth that ideal space (*templum*) which the Augur formed in the sky with his staff. In spite of the lapse of so many centuries, and so many revolutions, all trace of this old Rome has not disappeared in our time. In different parts of the Palatine the remains of the walls, constructed by the first founders of the city, have been discovered, and may still be seen. These are large blocks of stone, drawn from the hill itself, and on which, later on, the Cæsars rested the foundations of their palaces. When the imperial palaces fell, the old remains covered by them were brought to light again. Not only have we here and there identified the boundary of primitive Rome, but its chief entrance is believed to have been found. Towards the Arch of Titus a road leaves the Sacred Way and wends straight up the hill. It is neither broader nor less steep than the rest, and is only distinguished from all the others known to us by the large size of the slabs forming its pavement. This was the Palatine road or ascent (*clivus palatinus*).[1] Scarcely do we enter upon it, when the still visible supports of a large gate are met with. A little further on, enormous blocks of stone detached from a wall have rolled to the ground. The wall was the very one attributed to Romulus; the gate is much less ancient, but it is believed to have replaced that which served as chief entrance to *Roma quadrata*. It was called *Vetus porta* or *porta Mugonia*,[2] and is said to have received the

---

[1] See on the plan, No. 1.   [2] No. 2 on the plan.

latter name from the bellowing of the oxen which left it in the morning to pasture in the swamps that afterwards became the Forum. When the Emperors had established themselves on the Palatine, they had a new gate built, much finer than the first, which caused it to be forgotten. There were then no longer oxen nor swamps, but great lords and courtiers trod the Palatine road all day long on their way to see the master. It is, however, probable that the new gate was built quite close to the old one, and that the one shows us the site of the other.

But this is not the only discovery made on this spot. Whilst excavating to the right of the gate, they soon found a mass of large stones, in which it was not difficult to recognise the foundations of a very ancient temple. It can scarcely be doubted that this is the Temple of Jupiter Stator, one of the most celebrated of Rome, which, for want of a true site, archæologists have hitherto placed a little everywhere, according to their fancy. Titus Livius relates the occasion of its construction. The Sabines, after seizing the Capitol, had thence flung themselves on the soldiers of Romulus. The Romans fled dismayed. "Already," says the historian, "the army in disorder had reached the old gate of the Palatine, when Romulus, whom the fugitives had hitherto borne along with them, stopped, and raising his eyes towards Heaven: 'Jupiter,' said he, 'it was thou who didst encourage me to throw the foundations of my town upon this hill. I pray thee, father of the gods and of men, turn the enemy from us, calm my soldiers' fear, stop their shameful flight, and I will build thee here a temple which shall eternally recall to posterity that Rome was saved by thy help.'"[1]

It is the remains of this temple, dedicated to the god who stops all the flyers (*Jupiter Stator*), that have been found.[2] This point once fixed, the old town of Romulus is explored with tolerable ease. We have only to go over it in imagination, in order to find its chief monuments again. "Near Jupiter Stator," Titus Livius tells us, "lived Tarquin the Elder," and M. Rosa has placed a tablet on the spot where his house must have been. A little lower rose the Temple of Vesta, where the sacred fire burned, and its foundations are supposed to be still in existence under the church of *Santa Maria Liberatrice*. Behind St Theodore, on the slope of the hill opposite the *Forum boarium*, is the spot where, down to the last days of the Empire, the curious and the devout were shown a little grotto shaded by a fig-tree, called the Lupercal. It is here that the she-wolf was said to have suckled the divine twins, so a bronze she-wolf, the work of an Etruscan artist, was placed there, and is believed to be the one found at the beginning of the fifteenth century, which now adorns the Capitoline museum. A little further on, just where the church of St Anastasia stands, was the great altar (*Ara maxima*) said to have been consecrated by Evandor, where, down to the end of the Empire, the victory of Hercules over Cacus was celebrated.[3] Above, on the hill, was seen a monument

---

[1] Titus Livius, I. 2.　　　[2] See on the plan, No. 3.
[3] Servius, *Aen.*, VIII. 271: *Ara Herculis, sicut videmus hodie post januas circi maximi.*

more venerable yet, and which a true Roman could not visit without emotion. This was the house, or rather the cabin of Romulus, "where," says a poet, "two kings were content with a single hearth,"[1] and which formed a strange contrast with the marble palaces surrounding it. It was preserved and repaired with such care that it still existed at the end of the fourth century. Not only can we picture it to ourselves from the descriptions given of it by ancient writers, but a recent discovery has almost put it before our eyes. In excavating an old necropolis, near Alba, cinerary urns of terra-cotta were found, rudely worked, and representing a kind of small round edifice with a pointed roof. We know it to be the type of the ancient cottage of the Latin peasant, built of reeds and covered with straw. They were therefore accustomed to build their tombs in the image of their houses, and the abode of the dead was made like that of the living. The most ancient temples—those of Hercules Victor and Vesta—were also built on this model, and it was natural that the habitation of the kings should resemble that of the gods.[2] These monuments which used to cover the Palatine no longer exist, but we know where they must have been, and we run but slight risk of mistaking, when we recognise some of them in the ruins heaped together on different parts of the hill.

Perhaps it may be thought that I am treating these old memories too seriously, and that to appear to believe what Titus Livius or Denys of Halicarnassus tell

---

[1] Prop. IV. 1, 10 : *Unus erat fratrum maxima regna focus.*
[2] See on the *Casa Romuli*, Rossi, *Piante di Roma*, p. 1, *et seq.*

us of those remote times is to do them too much honour, but, as Ampère has already remarked, "if it is easy for a *savant* in his closet to laugh at Romulus and his successors; to see in the tales told us of them only extravagant fables, or explain them as myths devoid of reality, one does not feel quite the same assurance when one has just visited Rome. There, that past which at first seems so far off, so doubtful, draws near to us; it is touched and seen. It has left of itself such deep and vivid traces that it is impossible to deny it all belief. To be exact, had nothing remained of those ancient centuries, one might think that the Greek chroniclers who first unravelled the annals of Rome had amused themselves by inventing all sorts of fables, in order some way or other to fill up the gaps of history. But, impudent liars though we may suppose them, they were not free to imagine everything according to their caprice, since they found themselves face to face with memories they were bound to respect. These memories could not be lost, because they were attached to indestructible monuments as old as the very beginnings of the city. Generations passed the names of their founders from one to the other, and their sight recalled the disasters or the victories which had been the occasion of their construction. Annalists of the sixth century doubtless added much to these traditions. The imagination of the Romans was dry and short, and they lacked the art of embellishing their history with marvellous fictions like the Greeks. As time effaced the memory of the past, popular fancy did not know how to repair these losses by new and charming inventions. At the end of a few centuries, nothing remained of those ancient events but a few names and a few facts, on which it was easy to embroider a great many lies. But, if the lie covers all the surface, there must be a little truth beneath."

Such are the ideas inevitably suggested by a visit to the Palatine. They possess the mind with peculiar force when one happens upon those great ruins of wall I have spoken of, which formed the wall of Romulus. These walls are constructed on nearly the same system as those attributed to Servius, and can only be slightly anterior. Both are composed of blocks of tufa laid together, not united by cement, and kept in place by their weight alone. The arrangement of the layers is the same: the stones are placed alternately lengthways and endways. It is asserted that this manner of building belongs especially to the Etruscans, and that the Romans had it from them. Such was their usual system. "They took everywhere," says Pliny, "what they thought worth taking (*omnium utilitatum rapacissimi.*)" But if this sensible race borrowed without scruple of its neighbours, or even of its subjects, whatever could be useful to it, it knew how to adapt what it imitated to itself. In introducing inventions from outside, the Romans accommodated them to their peculiar genius; they took full possession of them, as it were, modifying and renewing them according to their wants: they were pupils who soon became masters. Beulé justly bids us remark that the Etruscans never produced great results in the grand art of building transmitted by them to the Romans, and that it reached much higher perfection at Rome than among themselves. The Romans gave it more and more their own character, and when they applied it to works of public utility, such as bridges, drains, or aqueducts, or to edifices especially admitting of grandeur and majesty, such as amphitheatres or arches of triumph, they evolved masterpieces from it. What shall I say? It seems to me enough to look on those fine walls, remaining to us from the royal epoch on the Palatine or elsewhere,[1] in order to foresee and divine the impulse which architecture is about to take in Rome, and in what direction its development will be. Those who built them, whoever they were, could not have been barbarians. Such great works suppose them to have reached a certain stage of civilization. They had at their command powerful means of placing the stones one upon the other, and raising them to such great heights. They were imbued with the consciousness of their worth, and with that confidence in their own duration which makes great peoples. They were not, like savages, content hurriedly to construct a temporary shelter to protect their sleep for a few nights against unforeseen attacks; they thought of the future, and worked for their descendants. In the midst of these swamps and these forests, they took care to raise defences which were to last thousands of years.

"They were already beginning," says Montesquieu, "to build the Eternal City." I add that they not only

---

[1] The finest ruins remaining of the walls of Servius are on the Aventine, opposite the church of St Prisca, in the *vigna Maccarani*, now belonging to Prince Torlonia. A fragment of wall is found there 30 mètres long and 10 high, wonderfully preserved, and which strikes one with surprise and admiration. All who desire to have an idea of these ancient buildings should go and see it without fail.

sought to make their walls solid. The manner in which those blocks are put together shows that they possessed, confusedly it may be, the instinct of grandeur, the feeling of proportion, and the taste for the kind of beauty which is born of strength. Assuredly, I repeat, they could not have been savages.

An important discovery, made not long since, proves how well founded are these conjectures. The important works undertaken in different quarters of the town since 1870, and, above all, those near the Baths of Diocletian, have resulted in the finding of many remains of these fine walls of the epoch of the kings. On examining them more nearly than had hitherto been done, it was perceived that signs were inscribed on these large blocks of stone. They are sometimes rather lightly graven, and then it is very difficult to read them, but the workman has in most cases traced a deep furrow, which has resisted time, and is as visible now as on the day on which it was cut. These were probably marks to indicate, sometimes the quarry whence the stones were taken, and sometimes the site for which they were destined. As they came from the neighbouring mountains, it was very necessary to inform those who conveyed them where they must be placed, in order to render any mistake impossible. These signs are very often letters, and most of those letters belong to the ancient Latin alphabet.[1]

This discovery, to tell the truth, was not quite a

---

[1] These characters were studied in a very interesting memoir of the learned Barnabite, Father Bruzza (*Ann. de l'inst. arch.*, 1876). See also the discussion of M. Jordan, *Topogr.* I. p. 259, *et seq.* ; Suétonius, *Vespas.*, 8.

surprise for the learned. Otfried Müller had indeed maintained that the first Romans did not know how to write, and that they only learned the art towards the time of the Decemvirs, when the twelve tables were promulgated; but M. Mommsen long since refuted this opinion. No doubt is any longer possible now that letters have been found graven on walls of the royal epoch. Hence it ceases to be absolutely improbable that written monuments of those remote ages should have been left. It was formerly the fashion to laugh at Suetonius, because he seriously relates that at the burning of the Capitol, under Vitellius, 3000 tables of brass perished, containing laws (*Senatus Consulta*) and plebiscites from the birth of the town (*pœne ab exordio urbis*).[1] People would not admit that, in the time of Augustus, there could still exist copies of the treaties concluded by Tullus Hostillius with the Sabines, and by Tarquin with the inhabitants of Gabii, although Horace says that they were the delight of antiquarians. Doubtless, it must not be too readily believed without proof that all these documents were authentic; but, after all, they may have been. We have at least no longer any right to disdainfully condemn the distinct testimony of historians like Denys of Halicarnassus, who affirmed that they existed, and that he had read them, without doing them the honour of discussion.

For it is now certain that the founders of Rome knew and practised the art of writing, and that they employed it for the ordinary uses of life. It was not

---

[1] Suetonius, *Vespas.*, 8.

with them the privilege of some classes of the nobles or the priests; it was used by the undertakers of public works, and perhaps even by the workmen. It would certainly be ridiculous to pretend with Cicero that in the time of Romulus science and literature already flourished at Rome, and picture to ourselves those senators, covered with the skins of beasts, as sages issued from the school of Pythagoras, and repeating its lessons; but it would be a still greater error to make them downright savages, barbarians ignorant of all knowledge and all the arts. Nor were they, on the other hand, quite epic heroes, as Niebuhr represents them. Ajaxes and Hectors came in a time when the exploits of warriors were only preserved in the songs of the rhapsodists; for such hypotheses of legends and epic recitals found little room at a period when reading and writing were known.

The city of Romulus was not destined to remain long enclosed within the narrow boundary traced for it by its first king. It soon overflowed on every side, and ended by occupying all the surrounding hills. Thenceforth the Palatine was no longer Rome itself, as at first, but it always remained one of the chief quarters of the enlarged town. Celebrated temples were found there in great number—that of Jupiter Victor; of the goddess *Viriplaca*, who reconciled households; that of the mother of the gods, whence every year, on the 27th March, started the merry train of devotees and begging priests, who went through the streets of Rome singing light songs, to bathe the statue of the goddess in the little river Almo. There, also, some of the most illustrious citizens fixed their abode. They liked to lodge as near as possible to the Forum and to places of public business. We know the exact position of the most illustrious of all these houses, that of Cicero, if it be true, as MM. Visconti and Lanciani think, that a large building, whose remains are seen at the corner of the Velabrum, belonged to the portico of Catullus. The house of Cicero, as we know, was quite close to it. He was very proud of living on the finest site in Rome (*in pulcherrimo urbis loco*). He tells us that he thence commanded the Forum, and that his view extended over all the quarters of the town. This house was associated with the vicissitudes of his fate. During his exile Clodius got the people to decree that it should be razed to the ground, and that in its place a temple to Minerva should be consecrated. After his return the Senate decided to rebuild it at the public expense, and Cicero obtained 2,000,000 of sesterces (400,000 francs) for its reconstruction. Does not this read like a narrative of contemporary history?

Of all those private houses constructed during the Republic, which sometimes recall such great memories, only a few ruins remain, and the preservation of even these we owe to a strange chance. Those placed on the top of the hill were demolished in order to make way for the dwellings of the Cæsars; but there were others situated in what is called by the barbarous term *intermontium* of the Palatine. The Palatine, like the Capitol, was originally divided into two by a narrow valley. It ran north and south, from the Arch of Titus to the grand Circus. This small valley was filled in by the emperors when they wished to extend and level the ground on which they were raising their palaces, and the houses that had been constructed there fell under the weight of the piled-up earth. Some, however, resisted, and the excavations have brought their ruins to light again.

In this connection it is as well that I should recall one of the reasons, the chief perhaps, which make the excavations at Rome always so prolific. This fecundity usually somewhat surprises those who are accustomed to our own modern towns, and to the process by which we see them renewed before our eyes. Rome, like all capitals, has in the course of its long existence been several times rebuilt; but the manner in which the Romans set to work to rejuvenate their town was less fatal than ours to the old ruins of the past. We now demolish; in those days they were content to bury them. We aim, above all things, at making straight avenues, and in order to facilitate circulation for the numberless vehicles which traverse our streets, we level the heights, we suppress the hills. It may then be said that the soil of Paris is continually being hollowed out; that of Rome, on the contrary, was constantly rising. The great Roman lords who wished to gladden their eyes with a more extended view, or who merely sought to enjoy purer air, in this pestilential climate, were accustomed to build their houses on immense foundations. Even when a new quarter was to be made, they began by filling in the old one with earth brought from elsewhere, and built upon that. One is pretty certain, then, in raising this earth, to find the primitive soil again, and the remains of ancient buildings.

This is what has happened on the Palatine, as everywhere,[1] and thus it is that under the palaces of the Cæsars some houses of an anterior date have been discovered. There is one especially called, I know not why, *the baths of Livia*, and of which some chambers still remain in a sufficiently good state of preservation.[2] Graceful ornaments are seen on its ceilings, groups, figures, arabesques thrown into relief by a gold ground, a whole set of decorations, at once sober and elegant, which give us a very admirable idea of Roman art under the Republic. The Palatine, about the time of Cicero and Cæsar, must have been filled with such houses; but this is the only one that has survived.

---

[1] They had the same good fortune in the excavations which were made a few years since at San Clemente. Their history is widely known; but I think it well to recall it, in order to show by a striking example what a wealth of discoveries may be expected in digging the soil of Rome. San Clemente is an admirable basilica of the twelfth century, containing fine frescoes of Masaccio. In carrying out some works there they brought to light, under the present basilica, a more ancient church, with curious paintings and columns of marble and granite. It went back to the time of Constantine, and had been used during seven centuries, down to the sack of Rome by Robert Guiscard. Encouraged by this success, they excavated more deeply, and were not long in finding, under the primitive church, a sanctuary of Mithras and some portions of a Roman house dating from the beginning of the Empire. Then, on going lower yet, constructions of tufa were discovered, certainly as old as the first years of the Republic, and perhaps even belonging to the period of the kings. Here, then, is a succession of monuments of all the epochs, and by descending a few steps one may have the spectacle of the entire history of Rome from its foundation down to the Renaissance.

[1] See No. 4 on the plan.

II.

THE HOUSE OF AUGUSTUS ON THE PALATINE — HOW, LITTLE BY LITTLE, IT BECAME A PALACE—WHAT REMAINS OF IT—EMPLOYMENT OF MARBLE IN THE IMPERIAL EPOCH—NEW PROCESSES IN THE ART OF BUILDING — THE PALACE OF TIBERIUS — THAT OF CALIGULA—THE CRYPTOPORTICUS WHERE CALIGULA PERISHED—THE HOUSE OF LIVIA AND ITS PAINTINGS—THE PALACE OF NERO.

WITH the Empire new destinies begin for the Palatine: it then becomes the dwellings of the Cæsars, and, as Tacitus expressed it, the centre of the Roman world (*arx imperii*). In his youth Augustus lived near the Forum; a little later on, when he was still only one of the ambitious ones who coveted the succession of the great Dictator, he bought a somewhat modest house upon the Palatine which had belonged to the poet Hortensius. It contained neither marbles nor mosaics, and was only ornamented with commonplace porticoes sustained by stone columns. Yet it was the beginning of those imperial palaces which, continually spreading, ended by covering all the hill. The house of Augustus grew little by little with its master, and it is not uninteresting to study the successive additions it received: for in the adroit manner in which, insensibly, and without shocking any one, he made the dwelling of a private individual into that of the chief of the State, I think we find the whole policy of this clever personage.

In seeking a secret reason for all his actions, one

runs no risk of rashness. Even in his most intimate
life it was his wont to leave nothing to chance, and he
is known to have written his conversations with his wife
beforehand, for fear of saying a little more than he
wished to do. We must therefore believe that if he
chose the Palatine above all the other quarters of Rome
for the location of his abode, he had some motive for so
doing, and these motives are not very difficult to find.
It was on the Palatine that the ancient kings of Rome
were said to have lived. Augustus was most anxious to
put himself into their company, and when he determined
to give up the name of Octavius, which the proscrip-
tions had brought into disrepute, and to take a new one
he was first tempted by that of Romulus, and he would
have preferred it to the others had not the violent end
of the first king seemed an evil omen for his successor·
It is certain, then, that in taking up his position on the
hill which had been the seat of royalty, he hoped to
inherit something of the respect by which those ancient
memories were surrounded. So he, as well as those who
came after him, took great care to preserve and repair
all that remained on the Palatine of that distant past.
It has been remarked that the imperial palaces often
respectfully turn aside from ancient ruins, the precautions taken not to include them in the new constructions being still visible. It was doubtless thought that
those venerable monuments of the ancient kings of
Rome protected and consecrated the habitation of the
new masters of the Empire.

Augustus also made a great point of doing nothing
hastily. His great art was to manage transitions with
caution, to avoid scandal and surprise in all things, and

to accomplish the gravest changes without noise. He
did not fail to proceed thus on the present occasion,
although it was apparently of less importance. He
knew that a monarch must have a palace, and that the
master of the world could not lodge like a private
individual. So he resolved to enlarge the house of
Hortensius, which was no longer adequate to his fortune.
After his victory over Sextus Pompey, when his power
was acknowledged by all Italy, which he had just
delivered from the fear of a servile war, he ordered his
stewards to purchase a certain number of houses surrounding his own, and demolish them. As these demoli-
tions might set suspicious minds thinking, he caused it
to be given out that he was not working for himself
alone, but in the public interest, and that he desired to
consecrate part of the ground to religious edifices.
And he really had the famous Temple of Apollo Palatinus
built, together with the Greek and Latin libraries, of
which mention is frequently made by writers of that time.
The magnificence of these buildings alone attracted public
attention, and it was scarcely noticed that at the same
time the house of the prince was also growing larger,
and changing its aspect. A short time afterwards the
new palace was destroyed by fire. It was customary in
Rome, after a misfortune of this kind, for the friends of
him who had been its victim to club together and
help him repair his losses, these voluntary con-
tributions standing in lieu of our insurances. The
fire on the Palatine was a natural occasion for showing
how many friends Augustus possessed. All the
citizens of Rome hastened to bring him their offerings;
but he would not accept them. He only took a trifling

sum, a denarius at most, per person, and rebuilt his house at his own expense—only he profited by the opportunity to rebuild it larger and more beautiful. When he was named Pontifex Maximus, instead of doing like his predecessors, and going to live in a special building near the Temple of Vesta, he remained in his own house and contented himself with raising a Temple of Vesta in it. Thus ancient usage seemed to be preserved, and the Pontifex Maximus was still near the divinity who protected Rome. In a curious and often-quoted passage, Ovid has taken pleasure in describing the house of Augustus for us, as it was towards the end of his reign. Exiled to the ends of the earth, and full of regrets for Rome, whither he was forbidden to return, the poor poet sent his verses to entreat for him. He represents them wandering in this city, where they have become strangers, obliged to ask their way of the passers-by, and seeking above all the dwelling of him who so cruelly punishes, but can also pardon. The directions they receive are so precise that we can still follow the road with them. There is first the Forum and the Sacred Way. "See," they are told, "here to the right is the gate of the Palatine, near the Temple of Jupiter Stator."[1] A little higher a house more beautiful than the others is seen, "and worthy of a god." It is surrounded by temples, adorned with arms and a shield; an oak crown shades its entry, and laurels are planted at the two sides of the door. Those laurels, that crown, solemnly awarded to Augustus by the Senate "in the name of the citizens he had saved," announced the dwelling of the world's master.

---

[1] Ovid, *Trist.*, III. 1.

The works of these late years have not yet restored the palace of Augustus to light, but the verses we have just cited tell us where to seek it. It was near the Temple of Jupiter, above the Palatine gate,—that is to say, at the spot covered by the gardens of the Villa Mills. Excavations were made there in 1775 by the Abbé Rancoureil, to whom the ground belonged, and under the ruins heaped up from everywhere, a two-storied house was found, whose arrangements were easily recognised. The upper storey had naturally much suffered, but the lower one was almost entire. Some of the rooms were filled with rubbish; others were empty, so that people could go through them and, what is more sad, could plunder them. They still kept their stucco, their precious pavements, and their marble linings attached to the walls by steel cramps. Charming pictures, much more delicate than those of Pompeii, adorned their ceilings. Admirable statues—among others the Apollo Sauroctonos, now in the Vatican—were found there intact. Great care was taken not to leave any object of art behind, from which any profit could be drawn. As for the remains of columns and pavements, they were carelessly removed, loaded on several carts, and sold in the lump to a marble dealer in the *Campo Vaccino*. The owner, a jealous amateur, as well as a skilful trafficker, kept his discovery as close as he could. He would not let other archæologists come near, and it is related that the celebrated Piranesi, who wished to see it, got into the garden by night like a thief, at the risk of being devoured by the dogs, and drew the ruins by moonlight. We still have the plan which he hurriedly made of them in the course of his adventurous excursion, and, what is better still, that of the architect Barberi, who carried out the excavations under the direction of Rancoureil.[1]

A glance at Barberi's plan suffices to show that this house, thought with much probability to be the palace of Augustus, resembled in its general arrangements all Roman houses. It contained an interior court, or peristyle, on which the different apartments of the palace opened. These apartments comprised a series of round, square, and rectangular rooms, corresponding pretty exactly with each other, in which the architect has apparently sought to unite variety with symmetry.[2] Two octagonal rooms, even, were found with forms so capricious that they recalled to those who saw them the fantastic constructions of Boronini. What at first caused some surprise,

---

[1] Barberi's plan was reproduced by Guattani in 1785, in the *Monumenti antichi inediti di Roma*, with very curious drawings of the chief monuments then found, and which were afterwards scattered or destroyed. A reduction of Barberi's plan will be found on our map of the Palatine. This part of the hill cannot now be seen; but it is said that it will shortly be opened to the public, and that the works destined definitely to give us back what remains of the palace of Augustus will ere long be resumed.

Among the rooms found in the house of Augustus is one wanting (as we know) in the chateau of Versailles, and which I will designate by its Latin name of *sterquilinium*. It is a genuine monument. Guattani, who asks leave to speak of it (*senza vergogna*) describes it in detail, and takes occasion to bid us remark *quanto gli antichi fossero ingegnosi nell' invenzione ed uso delle commodità le più indispensabili e necessarie all' umana vita*.

was to see that when these rooms and chambers are so numerous, they are generally very small, and that not one of them appears sufficiently extensive to serve for official receptions.[1] But Augustus, as we know, affected to live at home like an ordinary citizen. He wished to pass for a man, orderly, economical, and moderate in his tastes. He slept upon a low and hard bed; he only wore clothes woven by his wife or his daughter, he never had more than three dishes served at his table, and he takes great care to tell us in one of his letters that he sometimes fasted in the morning "more scrupulously than a Jew keeping Sabbath." There is, however, something of hypocrisy in this simplicity which he so complacently displays. Although he affected modest airs, his house inside, as we have just seen, was sumptuous. This prince, who always extolled ancient usages, nevertheless made a revolution in the manners and habits of his time. No one more than he assisted in the progress of the luxury he was in the habit of deploring. It is related that he caused to be read before the Senate and the people an old speech of Rutilius "against those who have the mania for building." He forgot that he had himself set the taste and the example by his magnificent constructions, and that a good part of the reproaches addressed by him to others recoiled upon himself.

"I found Rome of brick," he sometimes said, "and

---

[1] When Augustus, grown old, wished to assemble the Senate at a shorter distance from him, he convoked it in the Temple of Apollo Palatinus.— A. Gelle, Aug. 29th.

I leave it of marble." M. Jordan rightly bids us remark that never was metaphor more a truth. Before Augustus, marble was seldom used in Roman buildings; under the Empire it came into general use. Not princes alone adorned their dwellings with it; it was seen at Pompeii even in the shops of fullers and wine merchants. But it is on the Palatine that it especially abounds. Nowhere is it found in such quantities, and one has some difficulty in imagining how the architects who built the palaces of the Cæsars could so easily manage to get these rare and precious marbles, which came from all parts of the world, if a discovery made some years ago did not help one to understand. On the banks of the Tiber, not far from that strange *Monte Testaccio*, formed from the sherds of broken vases, was found in 1867 an ancient Roman harbour. The rings which attached the vessels to the stone quay, the steps by which the freight was lowered or raised, are still visible. Around the harbour large magazines were built, in which merchandise was temporarily piled after its disembarkation. When first discovered these still contained a large number of blocks of marble, on which the hewing process had been commenced. The inscriptions graven on these blocks, like those on the stones of the old walls of Servius, give us curious information concerning their origin and the manner in which they were brought to Rome.[1] The most celebrated quarries in the entire world, those which produced the most

---

[1] It is again the indefatigable Padre Bruzzi who has gathered these inscriptions and explained them in his memoir, entitled *Inscrizioni dei marmi grezzi*.

renowned marbles, belonged to the Emperors, who reserved them for the monuments they caused to be erected. The works undertaken and the number of workmen that had to be employed on them, became so considerable under Trajan, that a special administration (*ratio marmorum*) had to be formed which was doubtless dependent on that of the privy domain (*ratio patrimonii*). Each quarry was directed by a steward of the Emperor (*procurator Cæsaris*), who had under his orders all kinds of officials—secretaries, superintendents, and artists. The workmen were very numerous, and in great part composed of people condemned to the mines by the tribunals of the Empire. These unfortunates, as a rule but little fitted for such rough labours, came to bury themselves alive in these detested caverns under the hard rule of slaves or freedmen. It was one of the severest punishments a judge could inflict, and during the persecutions it was very often applied to the Christians. To have hewed the marble from the quarry was not all—it had to be brought to Rome. From the ports of Greece and of Asia, of Alexandria and of Carthage, heavy ships were always starting, laden with enormous blocks, which crossed the sea with infinite trouble and exposed to all kinds of dangers. As large vessels could not ascend the Tiber, they unloaded at Ostia, so the Government had established an entire administration there, charged to receive the marbles and forward them to Rome. Medium-sized blocks were placed on barques of ordinary dimensions; but for monolith columns, colossal statues, or granite obelisks, special craft had to be constructed. Let the reader imagine the expense involved in these complicated operations, and the money that must have been paid to these thousands of workmen, officials, and sailors. Let him picture to himself what the marble cost from the day it left the quarry to that on which it was brought to the studio of the artist who was to cut it. But it was necessary to strike the eye of the crowd, and always give it fresh marvels to admire; it was indispensable that the *public happiness*, of which mention is so often made in inscriptions and in medals, should shine in the sight of all. In order that people might not be tempted to accuse the decrees of the Senate of mendacity—which at the accession of each prince celebrated the re-establishment of prosperity and the assurance of the peace of the Empire—in order to give manifest proofs of this prosperity festivals had unceasingly to be increased and monuments multiplied. And thus, from Augustus downward, magnificence became a political institution and a means of governing the world.

This policy was singularly favoured by fortunate circumstances: at the very moment when the princes plunged into these magnificent constructions in order to busy and dazzle the people, a kind of revolution was being accomplished in the art of building, which rendered their prodigality more easy. During several centuries, M. Choisy tells us in his learned work,[1] the Romans had used for their monuments enormous blocks of stone, rough or cut, but always placed one upon the other without cement. They never quite

---

[1] *L'art de batir chez les Romains*, par A. Choisy, Paris, 1873.

abandoned this mode of construction, in which each stone awoke the idea of a difficulty overcome, and gave the entire edifice an air of power and grandeur; but as it was slow and costly, they of the Empire preferred another. Instead of composing the body of their monuments of large blocks painfully heaped up, they got into the habit of using irregular materials, put together in fragments and joined by mortar. This process, which they doubtless did not invent—I said just now that they rarely invented—but of which they first made a general and methodical use, offered wonderful advantages to people who wished to build cheaply and at small expense. "It allowed of their raising colossal vaults with the help of bricklayers alone, and without other materials than lime and pebbles." The source whence they had it, and the period of trials and gropings in the dark which they passed through ere they learned how to use it, are not now known to us. M. Choisy bids us remark that the Pantheon is one of the most ancient, and, at the same time, the finest of the monuments built on this system. It was, therefore, under Augustus that it reached its perfection. It was so conformable to the practical sense of the Romans, and so useful to their policy, that it lasted throughout the duration of the Empire. "Amidst the general decline of the arts," says M. Choisy, "the good traditions of Roman building were perpetuated without alteration, and without progress. Under the Antonines, they did not build otherwise than under the Cæsars." It was the employment of these economical and rapid processes, which were successfully used until almost the last day of Rome, that rendered the great constructions of the Empire possible.

Tiberius was not so prodigal as Augustus, nor was he so fond of building, yet there are some mementoes of him on the Palatine. It seems that he did not inhabit his predecessor's house, but had his palace apart, which was called by his name (*Domus Tiberiana*). It is several times mentioned in the accounts of the historians, and what they tell us makes the spot where it was situated known to us. Among these narratives there are some which it is impossible to forget. Tacitus relates that, on the 15th January 69, the Emperor Galba was making a sacrifice in the Temple of Apollo, near the palace of Augustus. He had at his side one of his friends Otho, who coveted the Empire. The gods seemed adverse; the signs observed in the entrails of the victims were unfavourable; and an aruspice announced to the emperor an imminent peril. Otho rejoiced; for he was not unaware that the moment approached when the conspiracy hatched by his friends against the old emperor must break out. Suddenly one of his freedmen comes for him, and, on a word agreed upon, takes him away with him. Leaning upon his arm, Otho traverses " the house of Tiberius," descends thence on to the Velabrum, and, turning to the right towards the Forum, arrives near the Temple of Saturn, about the Golden Milestone, whence started all the roads of the Empire. There he meets twenty-three soldiers of the prætorian guard, who proclaim him emperor, throw him into a litter, and take him to the camp, "while Galba," says Tacitus, "continued to weary with his prayers the gods of an Empire that no longer belonged to him."[1] The house of Tiberius must therefore have been placed to the north of the Palatine on the same side as the Velabrum. It was probably an old dwelling of his family, which he had caused to be enlarged, in order to put it on a level with his new fortune. Only a few small chambers now remain of it, which must have been the lodgings of slaves. Perhaps more will be found when the gardens still covering the ancient buildings have been excavated.

The palace of Caligula was a little higher, near the angle of the Palatine, looking towards the Forum. It is said to have been sumptuous, and that it was adorned with paintings and statues taken from all the famous temples of Greece. But the Palatine was not enough for Caligula; he pushed on his operations as far as the Forum, and turned the Temple of Castor into the vestibule of his house. By dint of hearing himself called a god, he had got to take his divinity seriously, and treated all the inhabitants of Olympus on a footing of equality. Not content with having had a temple raised for himself alone, where peacocks, parrots, and rare birds were sacrificed to him, he wished to take his share in the homage offered to the other gods, his colleagues. He often came to the Temple of Castor, seated himself gravely between the two Dioscuri, and thus yielded himself to the adoration of the nations. It is related that he one day saw among the crowd of devotees a shoemaker, who burst out laughing, and that he asked him, probably in order to give him an opportunity of repairing his fault, what effect he pro-

---

[1] Tacitus, *Hist.*, I. 27.

duced on him. "The effect of a great fool," replied the shoemaker. What is somewhat surprising is, that Caligula forgave him the boldness of his reply. But he one day got angry with Jupiter Capitolinus, the great Roman god, whom he doubtless accused of lack of consideration towards him. He was often seen, transported with fury, to murmur menacing words in the ear of the wooden statue. "One of us must disappear," he repeated to him, and it was feared that he would have the head of the venerable image cut off, and replace it with his own, as he had done to so many other gods, when he suddenly calmed down. "Jupiter," he said, "had begged his pardon;" and abruptly passing from fury to all the excesses of passionate affection, he would leave his new friend no more. In order to be nearer to him, and to be able to go and see him freely at every hour, he had a bold bridge built, which passed over the highest edifices of the Forum, and joined the Palatine to the Capitol.

This bridge was destroyed at an early date, and we have kept nothing of it; but the memory of Caligula is not therefore less lively on the Palatine. It remains attached to another ruin of the imperial dwelling which the excavations have given us back. Not far distant from the old Mugonia Gate, near the Temple of Jupiter Stator, there has been found one of those passages, called by the Romans *cryptoporticos*, which dived into the earth, and made it possible to pass from one habitation to another without crossing the streets or public places.[1] This is one of the longest known. It begins quite close to the *Clivus Palatinus*, skirts the houses of Tiberius,

---

[1] See No. 5 on the plan.

and Caligula for more than 100 mètres, and then,
turning abruptly to the right, continues to the spot
where it reached one of the palaces, now destroyed.
It must have been carefully decorated, and received
light from openings made in the roof. It is here,
in this doubtful light, that, on the 24th January of
the year 41, a terrible event occurred, of which the
historian Josephus has related all the details for us.[1]
Caligula was at first so loved by all the Romans that
they are said, in three months, to have immolated more
than 160,000 victims as thank-offerings to the gods
for his accession. But three years sufficed to make
himself feared and detested by the whole world; so a
conspiracy was formed, directed by the Military Tribune,
Cassius Cherea, to deliver the Empire. Cherea, although
no longer young, preserved certain habits of elegance in
his dress and of affectation in his speech—an air of
carelessness and softness—which made him thought
less energetic than he was. But under this foppish
appearance there was a soldier's soul; and he was,
further, a Republican who remembered the old Government
in the midst of a people eager to flatter the new.
Caligula, as insolent as he was cruel, ceased not to load
him with affronts. Every time the Tribune came, according
to custom, to ask for the watchword, the prince, in
order to ridicule his effeminate habits, delighted to give
him some low or obscene word, which made Cherea
the laughing-stock of the officers and soldiers. He
also seemed to pick him out for disagreeable employments.
One day he was ordered to examine
a female comedian, whose lover it was desired to ruin;

---

[1] Josephus, *Antiq. Jud.*, XIX. I. 15.

but the comedian, in spite of the most frightful sufferings, refused to say anything that could compromise him she loved. Cherea, displeased with himself and others, ashamed of the part he was made to play, and indignant at the affronts which he was forced to swallow, decided to slay the prince. After much hesitation, it was decided to carry out the project during the Palatine games that were given in honour of Augustus. These games were celebrated at the foot of the hill where, later on, the Arch of Titus rose. A temporary wooden theatre was built there, to which the crowd thronged during several days. That day it was more numerous than ever, for in the evening a strange spectacle was to be given—a representation of scenes in hell, by a troupe of Egyptians and Ethiopians. Towards noon the Emperor was in the habit of returning for a while to his palace to take a meal and rest. It is here that the conspirators awaited him. He left the theatre with his uncle Claudius and a few friends, preceded by the German soldiers who formed his usual guard. When he had passed the gate of the Palatine, he let his suite proceed by the road leading to the palace, and he himself turned aside to follow the *cryptoporticus*. He wished to see the children of noble birth, whom he had sent for from Asia for the games which he intended to give the people. They were employed in this retired spot in singing hymns and dancing the Pyrrhic dance. Cherea, who was the Tribune on duty, darted behind Caligula. He took care to send away the curious and the courtiers, saying that the Emperor wished to be alone, and followed him with the conspirators; then, approaching while he was talking to the young people, dealt him a sword-cut on the head. Caligula, who was only wounded, rose without saying anything, and endeavoured to flee. But he was at once surrounded by the accomplices of Cherea, who stabbed him thirty times with their daggers. The soldiers of the guard ran up at the noise, and the conspirators, who could no longer go back, because they would have met the Emperor's officers and the Germans coming to avenge him, continued to follow the passage as far as the spot where, according to Josephus, was the house of Germanicus, and by this way it was easy for them to escape.

The frightful tumult which followed the death of the Emperor should be read of in the historians. The Germans, who regretted him, killed all whom they found in their way round about the passage and the palace: the innocent and the guilty alike fell beneath their blows. Meanwhile the event began to be bruited in the theatre. No one dared to believe, although everybody wished it, and, says Suetonius, what well proves the terror under which people lived, is that it was imagined that the prince himself had set the news of his death going, in order to have the opportunity of punishing those who should seem to be glad of it. The strangest rumours circulated. People did not know what to do, and no one had the courage to display his feelings or to leave his place when the Germans arrived, more and more drunk with blood and anger, and seeing accomplices of the assassins everywhere, threatened to fling themselves upon the unarmed crowd. They were quieted with great difficulty, and the spectators fled in the midst of frightful disorder.

The *cryptoporticus* in which the tragic events occurred

is almost entirely preserved. It can be traversed in its whole length, and the imagination easily pictures to itself the terrible scene that passed in it eighteen centuries ago. We see again that prince, worn by excesses of all kinds, that old man of nine-and-twenty years, as Seneca and Tacitus have depicted him in indelible touches, with that little head upon an enormous body, those hollow eyes, that livid hue, that wild glance, that countenance which nature had made sinister, and which, by a strange coquetry, he took pleasure in rendering more terrible yet. We follow the assassins from the moment when they entered the passage with him, to that when they escaped by the house of Germanicus, seeking an asylum of the father after having killed the son. By a happy chance this house perhaps still exists, for some scholars think it the one found nearly intact at the extremity of the passage.[1]

It was discovered by Signor Rosa in 1869, and is certainly one of the most curious remains of the Palatine. There has been much discussion as to whom it could have belonged. Seeing it so close to the palace of Tiberius, it was natural to think it his family house, the one where he was born, and which his father bequeathed to him when dying. It was, indeed, the first name given to it; but some time afterwards there was discovered in the foundations a leaden water-pipe, on which was read at intervals the words: *Juliæ Augustæ*. This name, apparently that of the owner, was borne by many persons, and notably by Livia, the wife of Augustus, and M. Leon Renier is convinced that

---

[1] See No. 6 on plan.

it is she who is in question.¹ The house of the Palatine then, would be the one to which Livia retired on her husband's death. It is here, according to M. Renier, that she passed in sadness and isolation the last years of her life, an object of hatred and jealousy to her son, who blushed to owe his greatness to her. On the other hand, to MM. Visconti and Lanciani, our little house appears to be the one spoken of by Josephus, by which the murderers escaped, so they do not hesitate to call it "the house of Germanicus." However it may be with regard to these two opinions, which it would not be impossible to reconcile, and of which the second perhaps results from a false popular designation, the house is certainly much older than the passage, various details of its construction showing that it dates from the end of the Republic or the first years of the Empire. It continued to exist amid all the changes undergone by the Palatine. More and more hidden and buried by the great palaces that were built round about it, it had the good fortune to survive them. All the lower storey is in perfect preservation. Around the *atrium*, reached by descending a few steps, are arranged four rooms, still covered by the finest paintings, and the most intact yet discovered in Rome. Along the cornices run elegant arabesques, garlands of leaves and flowers intertwined with winged genii, and fantastic landscapes in charming taste. In the middle of the panels are

---

[1] M. Renier maintained this opinion in a memoir published by the *Revue Archéologique* in 1871, to which M. Georges Perrot has added an important study on the paintings of the Palatine since then. M. Perrot has reproduced the work of M. Renier and his own in his *Memoires d'Archéologie*.

seen five large frescoes forming distinct subjects. The two least important in size and merit are scenes of initiation and magic. Another, nearly 3 mètres high, represents a street of Rome, supposed to be seen through a window. This was a manner of enlarging an apartment or enlivening it, and of giving the Roman houses those street views which they generally lack. The custom still exists in our time. "All who have travelled in Italy," says M. Perrot, "know what a taste the Italians have preserved for those ocular deceits, for those perspectives which their decorators employ with a rare ability. You enter a court, and on the wall at the back, instead of the dull grey colour of the dirty plaster, or the glaring whiteness of the whitewash, you perceive a street receding in the distance, bordered by fine buildings or a garden; a coppice filled with birds flying about in the foliage; a trellis with ripe grapes hanging from it. The eye, without being cheated, still feels a lively pleasure in this substitution, and the mind enjoys an illusion which may be more or less complete, according to the greater or lesser adroitness of the painter. From the artists who decorated the houses of the Campanian cities and of Imperial Rome, down to those in our days, who spread their distemper colours on the walls of the houses of Genoa, Milan, Padua, and Bologna, there is an unbroken tradition, a heritage faithfully passed on from century to century, through all political vicissitudes. The Palatine view represents a street, with houses on which at each storey there are seen, sometimes open terraces, and sometimes balconies, surmounted by a roof supported on columns, like a *loggia* of our days. Persons leaning at the windows look at the passers-by, and a woman has just come out of her door. As she is accompanied by a young girl who holds in her hand one of those dishes in which the sacred cakes were placed, it may be supposed that they are both going to make some offering in a neighbouring temple. It is therefore a real landscape, a corner of ancient Rome exactly reproduced, where we find what is wanting at Pompeii—many storied houses.

The two other pictures are mythological. In one of them Polyphemus is seen pursuing Galatea. The giant is half plunged in the waves, and to indicate that he is dominated by his passion, the painter has represented behind him a little wingless Cupid, standing on his shoulder, and holding him in leash with two ribands. Galatea, fleeing on a hippocampus, looks back towards the Cyclops. Her right hand rests on the croup of the sea horse, while the left, embracing its neck, supports a red mantle which falls to below the loins. The red drapery and the black mane of the hippocampus throw the whiteness of the nymph's flesh into high relief. In the background an arm of the sea is perceived, shut in by high bluffs. The hills are covered with trees, the water has kept its transparency. "I remember no ancient landscape," says M. Perrot, "in which a more happy and broader interpretation of Nature is found." The other fresco, the finest of all as regards its execution, represents Io at the moment when Hermes is about to deliver her from Argus. There can be nothing more elegant and more graceful than the attitude of the disconsolate maiden, with eyes turned skywards, and in the disorder of her grief scarce holding upon her breast a mantle on the point of escaping. Behind her, Hermes approaches in silence, hidden by a rock from the view of Io and her guard, while vigilant Argus takes not his eyes from his victim, and, all gathered together, seems ready to spring upon the dreaded liberator. " This picture," says one of the best judges of ancient painting, Professor Hellig, " reveals an extraordinarily skilful and sure hand. The outlines are very delicately graded and yet well defined. The colour scheme, pitched in relatively bright tones, produces an harmonious and eye-reposing expression. It would be difficult to find at Pompeii a figure equal to that of Io. On the Palatine the proportions are more slender and more delicate, the colouring more transparent, and softer, than with the Campanian painters. Must we explain this superior refinement of conception and execution by saying that the painters of Rome had many more opportunities of seeing and closely studying Greek originals than their brethren of the provinces? Must we, above all, think of the influence which must have been exercised on Roman artists by the realities that surrounded them, and by the elegance of the women of the world in the great city? This I do not dare to decide." [1]

It seems very surprising that this elegant house, scarcely separated from the imperial palaces by passages and streets, should have been able to subsist without notable changes from the end of the Republic

---

[1] At the Paris *École des beaux arts*, we have a very exact copy of these paintings, the work of M. Leyrand, an inmate of the *Académie de France* at Rome. These pictures are placed in the vestibule of the hall where public exhibitions take place, on the side towards the *quai Malaquais*.

until the ruin of the Empire. Perhaps it was protected by the memory of illustrious dwellers who inhabited it during its first years; perhaps, too, the Cæsars who followed had a private reason for keeping it and repairing it with such care.[1] Whatever pleasure one may find in being an emperor or a king, there are moments when this enslaving vocation wearies, and one feels a want to descend a little from the heights. This official and public life would jade the most intrepid of the ambitious, were it not varied from time to time by a little solitude and shade. Louis XIV. himself, so made for this perpetual stage-playing, and accustomed to it from his youth, went to Marly, where etiquette was less strict, in order to escape from what Saint Simon calls "the mechanism of the court," and belong a little to himself. Who knows if this charming little house, so near the imperial palaces and yet so independent of them, where nothing recalls the supreme dignity, did not at times serve the princes, tried with the cares of Empire, as a retreat? It was quite adapted to their relaxation; it offered them a picture of private life, to which we always turn with some regret when we have left it. To my mind, setting aside the pleasure produced by the beautiful paintings which cover the walls, the thought that princes like Vespasian or Titus, Trajan or Marcus Aurelius, often frequented it, passing pleasant hours there in sweet chats with their friends, augments the interest we feel in visiting it.

Of Nero nothing remains on the Palatine. The taste

---

[1] The inscriptions on leaden pipes found there prove that it was repaired under Domitian and under Septimus Severus.

for the gigantic being predominant in him, his dream was to make a palace in which the whole town should be contained. The narrow hill, already covered with temples and houses that must be respected, afforded him no room for the edifices which he meditated, so he resolved to build his palace elsewhere. Caligula, in order to construct his, had already encroached upon the Forum. It occurred to Nero to equal the gardens of Mæcenas across the broad plain separating the Palatine and the Cœlian from the Esquiline. When the terrible ten days' conflagration had cleared the ground of the houses that encumbered it, Severus and Celer, Nero's architects, set to work. Their bold imagination, rich in unexpected combinations, was made to charm a prince whose morbid mind only loved new spectacles and extraordinary conceptions. They built him a palace such as had never been seen. The immense space at their disposal was filled with buildings of every kind. At the entry, near the spot where Hadrian afterwards raised the Temple of Rome, they placed a statue of the prince, a colossus 120 feet in height, which was subsequently made into an image of the sun. Towards the Esquiline, where the ground is so fertile, vast meadows, fields, and vineyards were spread out, and woods where wild animals roved. In the centre of the plain a pond was dug, which, according to Suetonius, was as large as a sea, and on whose shores rose picturesque buildings. As for the palace, properly so called, it was all resplendent with precious metals and rare stones encrusted in the walls; so they named it the "Golden House." Immense porticoes were seen there; banqueting halls with ivory tables; water-jets pierced with

narrow holes, that spread upon the guests an impalpable
rain of perfumes and precious spices; and baths where,
in the reservoirs, sea-water and all kinds of sulphurous
waters were found in abundance. When Nero took
possession of his new dwelling, he deigned to thank his
architects, who had served him to his liking, and was
heard to say that at length he was lodged.

III.

THE FLAVII AND THEIR POLICY—DESCRIPTION OF DOM-
ITIAN'S PALACE—THE PALACE OF SEVERUS—THE
IMPERIAL BOX AT THE GREAT CIRCUS—LODGINGS
OF THE SOLDIERS AND SERVANTS.

THE dynasty of the Flavii, who replaced the Cæsars,
were bound to conduct themselves differently from
them. Their ennoblement being recent and as yet
unendowed with the authority that springs from ancient
memories, it was necessary to base it upon public
opinion, to listen to its complaints, and hold them in
great account. Of all Nero's insensate undertakings,
the building of the Golden House was perhaps that
which had most irritated honest people. It recalled one of
the most terrible calamities of his reign, the burning of
Rome, to which Nero was accused of having himself set
fire, in order the more easily to obtain the ground he
coveted. "The fire was scarcely extinguished," says an
historian, "when he hastened to use the ruins of his
country to build himself a rich palace. People were
indignant to see those fields, those gardens, those
meadows, which replaced so many poor houses, and, in

the midst of a town overflowing with inhabitants, all this immense space filled with a single dwelling." "Rome," it was said in malicious verses, "will soon be nothing but a palace. Prepare yourselves yet, oh citizens, to emigrate to Veii, unless Veii itself be included in the house of Cæsar."[1] Moreover, this magnificence cost very dearly : the Emperor's architects did not calculate; the treasury was always empty, and, in order to replenish it, recourse was had, as usual, to confiscations and assassinations, so that the Golden House seemed to recall all the crimes it had entailed. Not only did the new emperors take good care not to finish it—they destroyed it. The vast grounds which it occupied were in part restored to the public, and only that was kept which was necessary for the erection of some sumptuous monuments. On the site of Nero's ponds, the Flavian amphitheatre was built, now called the Colosseum. On the Esquiline the baths were begun, which afterwards took the name of Titus, and at the post of the *Via Palatina*, on the Sacred Way, an elegant arch of triumph recalled to mind the taking of Jerusalem. These monuments, by means of which the new dynasty endeavoured to make itself popular, had the advantage over those of Nero, that the people profited by them. "Rome," said a poet, "is put into possession of herself again, thanks to thee, Cæsar. That which was the pleasure of a single man, serves for the delight of all."[2]

So the Empire had returned to the Palatine, and

---

[1] Suetonius, *Nero*, 39.
[2] Martial, *De Spect.*, 2-12.

this time, to leave it no more. Vespasian and Titus practised the policy of Augustus, sparing no outlay for monuments destined for the public, while they themselves lived simply, rather like private persons than princes. They managed, it appears, with the old imperial palaces, which, since the fire, had been repaired; but this simplicity was not to the taste of their successor, Domitian. He had the mania, or, as Plutarch expresses it, the malady of building. Few princes have raised such magnificent edifices, and we are told that his palace was the finest of all. A man who caused himself to be worshipped, and who ordered that in petitions to him he should be addressed as "Master and God," could only dwell in a "sanctuary," for thus he himself called his house, and willed that it should be called. It was natural that he should endeavour to make himself a dwelling which should be worthy of such a name.

This palace, the admiration of contemporaries, has been completely brought to light by recent excavations. It is not quite a discovery, for towards the beginning of the last century, Francis I., Duke of Parma, who possessed this part of the hill, had it excavated by the learned Bianchini, when a considerable quantity of ruins were found, and it was agreed, without hesitation, that they must belong to the palace of Domitian. It was then in a much better condition than now, and several rooms had preserved important remains of their primitive decoration. After everything that could be carried off had been taken to adorn the museums of the Farnese family, the ruins were covered up again with earth, and remained so for a century and a half.

Signor Rosa has given them back to us definitely; and, as they have this time been more completely cleared and disencumbered, as the general plan of the edifice is easy to reconstruct, and as it seems better to correspond to the idea we form of a palace, it is also the spot on the Palatine which strangers prefer to visit, and of which they keep the best recollection.[1]

Domitian's palace is still a Roman house, built on the same plan as the others—yet with the difference that its proportions are vaster. It is reached by the steep incline (*Clivus Palatinus*) which, as I have said, leaves the Sacred Way near the Arch of Titus, and served the Romans as their usual entrance to the Palatine, from the time of Romulus. At the end of this road was the principal façade of the palace. Under a magnificent portico, raised on columns whose shafts have been found, three doors opened. That in the middle gave access to one of the largest and boldest rooms known to us. It was doubtless the reception hall, and Signor Rosa has retained its ancient name of *tablinum*. In it the prince gave his audiences; it is there that he received the ambassadors of kings or of foreign peoples, and the deputations from the provinces which, on every anniversary, came to bring him the felicitations and good wishes of his most distant subjects. This hall is a living witness to the progress made by monarchical usages since the time of Augustus. At its end, opposite

---

[1] M. Ferdinand Dutert, who studied these ruins while in course of discovery, made a restoration essay, of which he published a summary in the *Revue archéologique* of January and February 1873. I am indebted to his kindness for a photographic proof of his restoration, which I reproduce in the plate opposite this page.

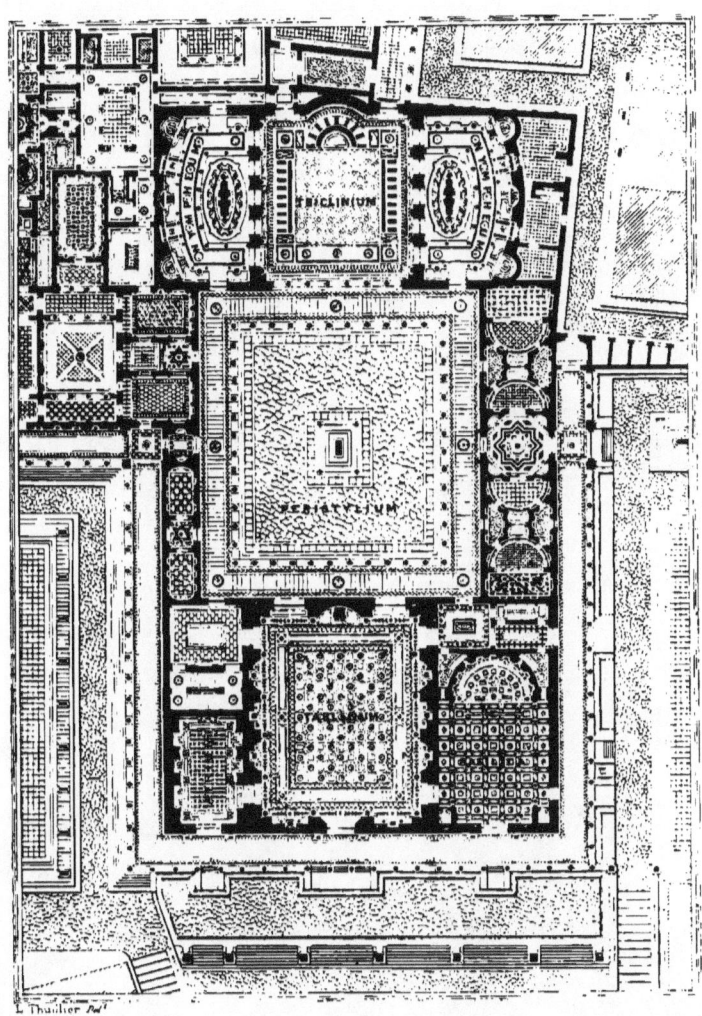

the entrance door, is seen a niche, doubtless destined to contain the Emperor's throne—for Domitian *had* a throne, and it was he who introduced into the imperial court the etiquette of Oriental monarchies. Statius, his favourite poet, openly gave him the name of king, which Cæsar had not dared to take, and he well knew that in applying it to him, he did not risk his displeasure. The decoration of the hall was in keeping with its extent. Bianchini relates that, when he discovered it, he found admirable remains of its ancient splendour. Around walls, covered with the most precious marbles, rose sixteen Corinthian columns, twenty-eight feet in height, and marvellously worked. Eight large niches, surmounted by four pediments, like those of the Pantheon of Agrippa, contained eight colossal statues in basalt, two of which, a Bacchus and a Hercules, were found in their places. The entrance door was flanked by two columns of *giallo antico*, which were sold for 2000 sequins (£925). The threshold was formed of such an enormous piece of Greek marble that it was converted into the table of the high altar of a church. All these riches have been dispersed; along the walls and on the pavements, there remain but a few fragments of the marble which covered them, and these relics no longer suffice to give us an idea of what must have been the magnificence of the hall.

The *tablinum* is placed between two other rooms of unequal size, opening, like itself, upon the entrance portico. The smaller of these was thought to be one of those household chapels where the divinities of the family were worshipped, and it has been named the *Lararium*, but whether this was its destination is somewhat uncertain. With regard to the other, however, there can be no doubt. It was a basilica, that is to say, one of those halls in which justice was administered. All its parts are clearly distinguished, and near the semicircular niche where the judges sat, there even remains a fragment of the balustrade which separated them from the bystanders. Here it was that the emperor judged the civil or criminal causes that were submitted to him. Domitian was very tenacious of this prerogative of his supreme power. He desired to assume the reputation of being a severe judge, and pitilessly chastised in others all the faults he so easily pardoned in himself.

Behind these three halls, which take up the front of the palace, is the peristyle, a vast court of more than 3000 square mètres in extent, surrounded by porticoes.[1] The remains of the fluted columns of Carian marble, which supported the roof, and the slabs of Numidian marble which covered the walls, are still seen. At the end of the peristyle, facing the *tablinum*, a wide door leads to the *triclinium*, or refectory of the palace. Martial tells us that, before Domitian's time, the Palatine had no *triclinium* worthy of the Cæsars, and congratulates him on having built one, which to the poet appears as beautiful as the banqueting hall of Olympus. He declares "that the gods might drink nectar there, and receive from the hands of Ganymede the sacred cup." This comparison is bold, yet it must be owned that the hall must have

---

[1] It was not possible to clear it all. A strip of ground imbedded under the terraces of the Villa Mills yet remains.

been very fine when it was intact. According to
Roman usage, it contained three tables. Two of these
were ranged along the lateral walls, and the chief
one was placed opposite the door of entry, in a kind
of niche, magnificently decorated, which still preserves a
portion of its pavement of porphyry, serpentine, and
giallo antico. This was the one where sat the emperor
and the greatest personages. The middle remained free
for the service. On each side, five large windows,
separated by columns of red granite, opened on two
nymphæa, in the middle of which the remains of a marble
basin are still found, ornamented with little niches
which must have contained statues. From the couches
on which the guests reclined for the repast, they
could see the water gushing from the fountain, and
falling in a cascade from step to step in the midst
of verdure, marble, and flowers. This elegant dining-
room is often mentioned by the writers of the time.
Domitian, who piqued himself on a love for letters,
and who, in his youth, had written verses which his
flatterers found divine, sometimes deigned to invite
poets to his table. Statius, who obtained this envied
honour, has described his joy to us in one of his *Silves*.
It is a veritable delirium. He declares that on
entering the Emperor's *triclinium*, he thought himself
transported into the midst of the stars, and that he
seemed to take his place at the table of Jupiter
himself. "Is it indeed you whom I see," he says to
the prince, "you, the vanquisher and father of the
world subdued; you, the hope of men, and the care
of the gods? So I am near you! In the midst of
the goblets and the viands which cover the table, I

contemplate your countenance." And he hastens to add: "I own that all the sumptuous appurtenances of the repast, these tables of oak supported on ivory columns, this army of slaves, allure not my looks. It is the Emperor alone I wished to see, and him alone I contemplate. I could not take my eyes from that calm countenance, which seemed to wish to temper the brilliancy of his fortune in an air of serene majesty. But he could not hide his greatness; it shone in spite of him, upon his features. On seeing him, the most distant nations, the most barbarous hordes, would have recognised their master."[1] These compliments appear rather strong when we consider that Domitian was in question, but the honour which the prince had done Statius was of those which turned the heads of poets. Martial declares that if Jupiter and Domitian were to invite him to dinner on the same day, he would leave the master of the gods, and go to the Emperor.

Of all these great halls we have now only marble pavements, bases of columns, and a few fragments of wall; the rest is destroyed. But the testimony of contemporary authors is sufficient to give us an idea of what we have lost. They are unanimous in celebrating the vast proportions of the edifice and describing its height. They say, in their hyperbolical language, "that in beholding it, one would think to see Pelion on Ossa; that its roofs pierce the ether and see Olympus more closely; that, from below, the eyes can hardly distinguish the roof, and that the golden pinnacle

---

[1] Statius, *Silv.*, IV. 2.

confounds itself with the beaming brightness of the skies." They tell us of that infinite number of columns "that might uphold the celestial vault while Atlas reposes for a moment;" they enumerate the marbles of all kinds that entered into the decoration of the walls, and they even persist with so much complaisance in these pompous descriptions that, against their intention, the thought occurs to us that there must be something of profusion and excess in all these ornaments. Simplicity was no longer loved in Domitian's time. The public taste and artistic talent had become less sure. Men no longer knew how to make beautiful, and sought to make rich. This is usual with all decadent arts. The prince, especially, was a passionate lover of this licentious magnificence, and a joker compared him to King Midas, who turned all he touched into gold.[1]

This immense palace contains many other rooms less important than those we have just described, but all the apartments necessary to private life have not been discovered. So it only served for official representations, and in reality the princes lived elsewhere.

---

[1] In several MSS. of the Middle Ages, the description of a palace is found, in which Signor Rosa has recognised the palace of Domitian (*Piante di Roma*, p. 123). This curious fragment shows, first of all, that the names given to the various rooms composing the palace are exact. They are found again in the Middle Age description. The reception hall is called *salutatorium*; it has beside it the *consistorium*, that is to say, the basilica, and further on the *trichorum*, or three-bedded dining-room (*triclinium*). It then shows that this beautiful palace subsisted until after the ruin of the Empire; that it was always the centre of the Palatine, and that it continued in the imagination of all as the type of an imperial palace.

Their true dwelling seems at all times to have been either the house of Augustus or that of Tiberius. In order to be able to pass from the latter to Domitian's palace without traversing the street, a subterranean gallery had been excavated, which still exists, communicating with the *cryptoporticus* of which I have spoken.[1] Thus the life of the emperors was, so to say, divided into two parts. The first, and doubtless the less agreeable portion, they spent in this magnificent palace, on whose door the emperor had inscribed *Aedes publicæ,* in order to make it understood that everybody had the right to come there and demand justice; the rest of their time they had lived in an abode less sumptuous, but more retired, more convenient, more fitted for family life, where, after having accomplished their business as emperors, they could, to use Antonine's fine expression, "enjoy the pleasure of being men."

The Cæsars had been living for a century—the best century of the Empire—in the old palaces, when it occurred to Septimus Severus to build a new one. Perhaps the opportunity was afforded him by the terrible fire which devastated the Palatine at the end of the reign of Commodus; but he certainly had another reason for doing it. New dynasties always feel the need of striking the imagination of nations by some great enterprise. This one, especially, which followed the Antonines, and which had to earn forgiveness for its foreign origin, affected to concern itself greatly with Rome, its adornment and embel-

---

[1] See No. 7 on the general plan of the Palatine.

lishment. Severus, like all who suddenly attain to high fortune, was always in fear lest his former situation should be recollected, and wished to obliterate its memory. It is related that when he returned to his country invested with a public function, one of his old friends, happy to see him, having embraced him, he had him beaten with rods, in order to teach him to treat a magistrate of the Roman people more ceremoniously. It doubtless appeared to him that, in rivalling his predecessors in magnificence, he showed himself worthy to succeed them. He desired to take possession of the imperial hill, by building there a palace which should bear his family name.

The Palatine was beginning to be encumbered, and room for new constructions must have become scarce. Yet a space still remained free opposite the Cœlian, along the Triumphal Way. It had been less built upon than elsewhere, because it descends towards the plain by gentle slopes, and does not afford level ground on which a vast edifice could be erected. Yet Domitian's palace had somehow or other extended thither. From that peristyle of which I have spoken, and which covered so large a space, a series of rooms, still but little known, communicated with the house of Augustus, which Domitian had thus included in his vast palace. Beyond the house of Augustus he had constructed a *stadium*, now entirely cleared. By the word *stadium*, was designated a kind of circus destined for foot races or athletic games. This was one of the favourite amusements of the Greeks. Nothing pleased that nation of artists more than to see a fine body display in varied exercises its strength and grace. The Romans, who were only struck by the indecency and danger of these exercises, did not like them. They imbibed a taste for them, however, under the Empire, and it was Domitian, especially, who worked their acceptance. He constructed for these games a large circus in the Field of Mars, of which the *Piazza Navona* still preserves the form and plan. He liked to preside at them, clad in a Greek costume, his shoulders covered with a purple mantle, and a crown of gold upon his head. It is not surprising then that he should have chosen to have in his palace a *stadium* where he could give for himself and his friends alone the entertainment which in the Field of Mars he shared with all the Romans. He doubtless liked to try, in the company of a few connoisseurs, the rapid runner or the skilful athlete whom he was afterwards to show to the people. The place where he gave these entertainments must have been very elegant;[1] the imperial hemicycle has been found, composed of two rooms, one above the other, of which the highest seems also to have been the most beautiful.[2] All around the circus were two tiers of porticoes supported on marble columns. The aspect which the place presented when the Emperor was seated in his box, and the courtiers, happy to take their part in these imperial distractions, crowded beneath the porticoes, may be imagined.

---

[1] In the library of the *Ecole des beaux arts* there is a very interesting restoration essay of Domitian's *stadium* (*Essai de restauration du stade de Domitian*) by M. Pascal, late inmate of the Académie de France, at Rome.

[2] See No. 8 on plan.

It was beyond Domitian's *stadium*, at the very angle of the hill, towards the east and south, that Severus built his palace. The expense of it must have been very considerable. Before constructing the palace itself, it was necessary, so to speak, to make the foundation on which it was to rise. We saw just now that the ground slopes gently downward towards the plain. They raised it by means of immense substructions, consisting of stone arcades superposed. These substructions still exist. The earth that covered them having disappeared, the arcades are seen on all sides, mounting one above the other, and forming strange groups. They are so high, and strike with such astonishment him who beholds them from the neighbouring roads, that people sometimes do them the honour to take them for the palace of the emperor itself. They are only its foundations and underground supports, however; the palace of Severus was built above them. A few still solid walls yet remain of it, the highest and best preserved found on the Palatine. One of them supported a magnificent staircase leading to the upper stories. But of all these imposing ruins, nothing equals in interest what remains of the imperial box on the Great Circus.[1] It was contiguous to the palace itself, so that the Emperor might be present at the chariot and horse races without leaving home. It consisted of a closed room, in which the Emperor and his family could take a little rest, and of a terrace whence the eye embraced the entire circus. The view enjoyed from it on a day when one of those great festivals was held, which brought together all the

---

[1] See No. 9 on the plan.

Roman people, must have been admirable. The long and close valley, extending between the Palatine and the Aventine, is to-day one of the saddest and poorest quarters of Rome. It was then an immense hippodrome, adorned with columns, obelisks, and statues, and surrounded by rows of marble seats, on which, during the public games, nearly 400,000 spectators crowded. Nothing could equal the animation of this crowd when horses or charioteers beloved of the public were to run. "The spectators," says Lactantius, "formed the strangest of spectacles. They were seen to follow with passion all the incidents of the race, to gesticulate, to cry, to howl, to jump upon their seats. Each of them took the part of one of the different factions. They insulted, they applauded the drivers, clad in green or blue, white or red, who turned about the *spina*. From the moment when the magistrate who presided at the festival gave the signal for starting by throwing a white handkerchief into the arena, to that when the most fortunate chariot, after traversing a distance of seven kilomètres and a half, touched the goal, a terrible noise which is said to have been heard at several leagues from Rome, arose from all these spectators. The Emperors shared the general excitement. They also had their favourite horses and charioteers, and did not willingly accept their defeat. I imagine that it was here in this imperial box, which a kind fate has preserved for us, that the strange scene took place related by Herodotus. Some of the spectators having ventured to hiss a driver of the blue faction, favoured by Caracalla, he ordered his guards to punish the culprits. The soldiers threw themselves on the seats of the circus, and to save themselves the trouble of picking out the offenders, slew all whom they could reach. It was an indescribable scene of confusion and slaughter, at which the Emperor, who saw everything from his box, must have been much edified.[1]

Septimus Severus is the last of the emperors who had a new dwelling built. After him the Empire became too wretched for a prince to be able to allow himself such a luxury. I have therefore finished enumerating the palaces that were built on the Palatine; but it contained other edifices besides the dwellings of the emperors. Near the prince his guards and his servants had to be lodged. Although these houses of soldiers and slaves were necessarily constructed with less care and at less cost, traces of them, nevertheless,

---

[1] Another part of the palace of Severus continued very famous. At the bottom of the hill, facing the Cœlian, he caused to be built along the Triumphal Way a three-storied portico, named the *Septizonium*. He desired to make it the chief entrance to the palace, but the Præfect of Rome, who was doubtless tenacious of ancient customs, prevented him, by causing the statue of the emperor to be placed at the spot where the door should have been. The *Septizonium*, therefore, was no longer anything but a magnificent monument, serving no purpose. The malicious who saw it placed opposite the road from Africa, pretended that Severus, in constructing it, had wished to strike his countrymen with admiration on their arrival. The *Septizonium* had the good fortune to pass through the whole of the Middle Ages without much accident. It was still nearly intact when it pleased Sixtus V. to destroy it and use the columns for some church he was restoring. "The Renaissance of the Arts," says M. Dutert, "was the signal for the mutilation and dispersion of the finest artistic works." The Popes often destroyed ancient monuments which the Ostrogoths had repaired. Was it not Paul V. who destroyed the admirable remains of the Temple of Pallas, in the Forum of Nerva, in order to decorate the Pauline fountain? *Piu Goto de'Goti!*

remain on different portions of the hill. At the bottom of the Palatine road, near the Arch of Titus, the excavations brought to light a great number of rooms of unequal size.[1] Signor Rosa supposes that they were occupied by the Prætorian cohort which guarded the Cæsars, and it is indeed very natural to believe that the barracks were placed beside the principal entrance to the Palatine. It is hither, then, according to Tacitus, that the unfortunate Piso, just adopted by Galba, came at the first news of Otho's revolt, assembled the soldiers of the guard, and made them that honest, sad speech which was not destined to win the hearts of the Prætorians. But more curious than these formless ruins, whose destination, on the whole, is somewhat doubtful, are those found at the opposite extremity, towards the Velabrum. A road has been discovered there, almost entire, and pretty well preserved, in which what was called the Ascent of Victory (*Clivus Victoriæ*) is thought to have been discovered."[2] It is another relic of Rome of the earliest times. It was entered by the "Roman Gate," one of those whose origin was said to go back to Romulus. Thence a narrow steep way went towards the top of the hill. The road, bordered on each side by high houses, could never have been very light, but it must have been darker still after Caligula had it partly covered, in order to extend his terraces. The right side of this road, that supported by the hill, certainly belonged to the dependencies of the imperial palaces.

---

[1] See No. 10 on the plan. The ruins discovered on this spot seem to be of a very low epoch. MM. Visconti and Lanciani are tempted to believe that they belong to constructions of the Emperor Maximian.

[2] See No. 11 on the plan.

On entering the half-filled-up chambers which still exist, when the eye begins to get accustomed to the darkness, one is surprised to see that these sombre apartments, which at first seemed scarcely suitable even for slaves, are sometimes decorated with great elegance. Many have preserved their stuccos and mosaics; there are some whose walls are still decked with graceful paintings, and one of the balconies has kept its fine marble balustrade. If these houses, as it is natural to believe, were inhabited by the prince's servants, they must have been reserved for the most distinguished slaves and freedmen—for the aristocracy of the imperial domestics. There were doubtless among these people without country and without name, bought in the markets of Greece, some whose good graces were sought by the greatest lords, who dominated the Emperor, and who often governed the Empire. When they became important and rich, they submitted to live in these apartments without air and without light, in order not to be far from their master, as under Louis XIV., the most illustrious personages, possessors of great hotels and fine châteaux, crowded themselves into the tainted apartments of Versailles for the sake of always being before the eyes of the king. But if those slaves and freedmen felt themselves obliged to inhabit these dark chambers, they desired, as far as in them lay, to make them worthy to receive them. Such at least is the only way of explaining this luxury of paintings and marble, and this fine ornamentation lavished upon walls where they could scarcely be seen.

On the other side of the Palatine, near the Great Circus, one of those ancient houses has been found, which were preserved after the hill had been invaded by imperial palaces, and were devoted to the housing of the attendants.[1] It perhaps contained soldiers and slaves at different periods. The rooms around the atrium are full of those inscriptions, scratched or scribbled with charcoal, called by the Italians *graffiti*. They are for the most part the work of soldiers calling themselves veterans of the Emperor (*veteranus domini nostri*), and some are smart epigrams, in which the veteran complains of the small profit he has derived from his service.[2] There are others which seem to prove that at a certain epoch the school for young slaves (*pædagogium*) was established in this house, where were educated the children destined to serve the prince, to approach him, to bear him company, and cheer him with their conversation. Several of these children have left upon the walls inscriptions which seem to prove that the school did not greatly amuse them, and that they were glad to leave it. Here also was found the famous caricature, now in the Kircher Museum, of which so much has been said. It represents a man with an ass's head stretched upon a cross. Below, a parson, roughly drawn, raises his hand to his mouth and gazes upon the crucified one. The scene is explained by a Greek inscription, in which we read the following words: "Alexamenus adores his god." This is evidently a

---

[1] See No. 12 on the plan.
[2] On the wall of one of these chambers was seen a little ass turning the wheel of a mill. Below it was written the following legend:— "Work, little ass, as I myself have worked, 'twill do thee good" (*Labora aselle, quomodo ego laborari et proderit tibi*). This charming little drawing was recently destroyed by a storm.

pleasantry directed against a Christian. In the time of the Antonines it was believed, even in the most enlightened society, that both the Christians and the Jews worshipped an ass. A soldier or a slave of the Emperor Alexamenus, having embraced the new doctrine, was the object of his comrades' railleries, but he bore them with courage, and in the midst of this hostile world he did not deny his faith. In 1870 M. Visconti found an inscription, in which he makes profession of it in the following words, probably graven by himself:

ALEXAMENOS FIDELIS.

Although Christianity early made its way into the house of the Cæsars, this is the only memento of it remaining on the Palatine.

IV.

ASPECT OF THE HILL IN THE THIRD CENTURY—IT CONTAINS THE EDIFICES OF ALL TIMES — MONUMENTS OF THE IMPERIAL EPOCH — DIFFERENCES BETWEEN THE PALACES OF THEN AND NOW—BEAUTY OF THE WHOLE.

HOWEVER lengthy this study already, I think it useful to add to it yet a few more words. After having enumerated in detail the edifices which each century saw rise upon the Palatine, we must endeavour to form an idea of the effect that must have been produced by the whole. Let us then suppose ourselves in the third century, about the time when Septimus Severus had just built the last of all the imperial palaces, and let us imagine that in one of those moments, becoming more and more infrequent, when the Empire is calm and victorious, we pay a visit to the celebrated hill. At this moment it all belongs to the Cæsars; and their family, their soldiers, and their servants alone occupy it. It contains buildings of different ages, of which some go back to the very origin of Rome, but which are all kept up and repaired with the greatest care. No ruins sadden the eye; the Cæsars will suffer none anywhere. Nothing in their Empire must have an air of poverty and desolation to belie the prosperity of their rule. Is it not known that one of them went so far as to unceremoniously abolish the companies that had been formed to purchase the great domains, and which, after drawing a good profit from the lands by cutting them up, did not take the pains to keep the houses in order when they could not find buyers for them? The Emperor is indignant at this conduct, declaring in his edict that "this is a murderous commerce, inimical to the world's peace," and which insults the public happiness; that, instead of covering the fields with ruins, it behoves so fortunate an age to build new houses, in order to make the happiness of the human race the more shine forth."[1] Of course these maxims were to be practised on the Palatine more than elsewhere. It was fitting that all should be maintained in good order around the imperial palaces; so, in spite of the miseries of the Empire, nothing there was allowed to fall into ruin, and this is what explains how the most decayed old houses were preserved into the coming of the barbarians.

---

[1] This curious edict against "rings" among the Romans was published and commented on by M. Egger, in the *Memoires de la Société des antiquaires de France*, 4th Series, Vol. III.

So there were monuments of all ages upon the Palatine, and the great interest offered to a visitor was that within a restricted space it contained, as it were, the entire history of Rome. From the time "when the oxen of the Arcadian Evander came to repose there," down to that when the African and Oriental dynasty of Severus settled upon it, each century had left some memento. It held the dwelling of the first king and the palace of the first emperor; the spot was shown upon it, where lived the great consuls of the Republic, and the best of the princes. All the transformations of the national worship could be followed: the Temple of Jupiter Stator, that of Apollo, and that of the mother of the gods successively recalled the epochs when Rome was content with the divinities of Latium, when she admitted the gods of Greece, and, finally, when she sought the exalted creeds of the East, and prepared the way for Christianity. People came to visit all these movements, and the most ancient, although the most simple, were not made the least of. The Romans were not like those *parvenues* who blush at the humbleness of their origin, and seek to hide it. They found in it, on the contrary, a cause for pride, because it enabled them the better to measure the greatness of the way they had traversed. No period of their history was excluded from their gratitude; they knew that all the ages had worked for the glory of Rome; neither political hatreds nor party prejudices had power to make them unjust towards any one: however ardent disputes had been, time had appeased all, and nothing remained of the past but the ever-living memory of services rendered to the country. The patriotism of a Roman of the third century was composed of an equal admiration for the heroes of the Republic and for the great emperors, and he visited with the same feelings of respect and pride the cabin of Romulus, the house of Cicero, and the palace of Augustus.

That, however, which predominated, that which had left the most mementoes on the Palatine, was the imperial epoch. It is not quite exact to pretend, in accordance with the inscription of the Farnese Gardens, that it contained the palace of the Cæsars (*Palazzo de' Cesari*), which would lead one to believe that only one vast habitation existed there, unceasingly enlarged and embellished like the Tuileries by the new Emperors who came to inhabit it.[1] It was rather the palatial quarter. There were five different quarters which bore the names of the princes who had them built.[2] Nothing like it is seen in our modern capitals. In our days, when princes, from caprice or from vanity, chose to construct a new dwelling, it is always very far from the old one. Their desire is to change, and what they first seek are a different situation and new points of view. The two chief residences of the Popes, the Vatican and the Quirinal,

---

[1] In Bianchini's time, it was believed. The restoration of the imperial palace, as he imagines it, should be seen in his work, entitled *Palazzo de' Cesari*. It is an immense construction, externally somewhat like the Farnese Palace, where all is in keeping, and seems to be of the same time. Nothing less resembles the idea we now form of the Palatine.

[2] At least, that of Tiberius seems always to have kept its name. See *Aulu Gelle*, XIII. 19, and *Hist. Aug. Prob.*, 2.

are placed at the two extremities of Rome. Here, on the contrary, all is gathered upon the same hill. It had become the home of the Empire, and it seemed as though a prince could not reside elsewhere. Dion says that the places where the Emperors sojourned during their travels took the name of Palatine.[1]

This accumulation of palaces must have greatly impressed visitors. Let us imagine an intelligent and curious provincial, of whom in those times there were many—a Gaul, a Spaniard, an African—who came to see this Rome, of which all the world was talking. Even after the imperial Forums had been traversed, and the marvels of the Capitol admired, the Palatine was still fraught with things to excite his wonder. We can easily imagine the spectacle which met his eyes; for the excavations made of late years allow us to reconstruct the typography of the hill exactly. On arriving by that *Clivus Palatinus*, of which I have so often spoken, and passing under the old Gate of Romulus, near the Temple of Jupiter Stator, he had before him the façade of Domitian's palace. This palace, which first met his eyes, was also the most important of all, and that which seemed most in keeping with the majesty of the Cæsars. A space, believed to be the *Area Palatina*, situated to the right, divided the imperial palaces into two distinct groups. One of these groups included the houses of Tiberius and of Caligula, built to the north of the hill, along the Velabrum and the Forum, while the other group was composed of three different palaces, having their own façades, entrances,

---

[1] Dion, LIII. 1, 16.

and peculiar characteristics, which could communicate with each other, and, on certain solemn occasions, form a single palace. That of Domitian adjoined the house of Augustus, more retired towards the south, and occupying nearly the centre of the hill. In the same line, a little further on, was the palace of Severus, situated towards the southern angle of the Palatine. The remaining buildings, exclusive of temples and historical edifices, served for the housing of the Emperor's slaves and freedmen.

I am somewhat tempted to believe that if we could see the Palatine as it was in the third century, we should, although much admiring it, still make a few reservations. Our taste has acquired certain habits, and has assumed certain requirements that would not be entirely satisfied. The approaches to and environs of the imperial palaces would probably seem mean to us. The *Clivus Palatinus* is not broad, the *Clivus Victoriæ* is still more narrow, and the *Area Palatina* does not appear sufficiently vast. If Domitian's palace was as elevated as Statius pretends, we really scarcely know where we could place ourselves, in order to take in all its height. Inside, these magnificent dwellings would please us more. The halls, the courts, the porticoes, would excite our admiration. Yet I think we should be very much surprised not to find any gardens to them. When the Emperors wished to taste the pleasures of the fields, they went outside Rome. Quite close, on the Alban lake, and at Tibur, they possessed charming villas, which it was easy for them to visit when they chose. If they wished to enjoy the real country—the country rough and unadorned (*rus verum barbarumque*)—they went further. We know how happy Antonius was to gather the vintages of his great Latin domains. This sufficed them, and they seem never to have planted on the Palatine those luxurious gardens with which the rich of our days love to surround their houses.[1] Nero alone anticipated our tastes, but perhaps less from love of the fields than in order to give himself "the proud pleasure of forcing nature." It doubtless appeared to him extraordinary and well worthy of a Cæsar to bring woods into the midst of Rome and possess a pond of salt water ten leagues from the sea. These reservations made, we should, I think, be as much struck as the Romans by the beauty of the edifices built upon the Palatine. Although dating from different epochs, they could not have presented diversities offensive to a fastidious age. Fire—that chronic scourge of ancient Rome—had often overtaken them. Each time they were promptly rebuilt, for Rome, as Martial expressed it, was a Phœnix which grew younger by burning, and when they were raised again they were always harmonised a little with the fashion of the day. Thus inconsistencies, which might have shocked, had been effaced, and yet enough difference remained to attract by contrast the attention of visitors. Each of the

---

[1] Yet mention is made of the gardens of Adonis (*Adonea*) in Domitian's palace, but these must have been of very slight extent. By the word *Adonea*, the Syrians and Egyptians rather understood gardenets than real gardens. They were earthen vases in which, at the time of the feast of Adonis, plants were sown that grow and die in a few days. This hasty and brief vegetation was an image of the destiny of the hero whose premature death was celebrated.

palaces had its peculiar character and merits. That of Augustus must have been more simple and of a graver taste; that of Domitian sumptuous to profusion; that of Severus imbued with that relish for grandeur found again in the Baths of Caracalla. The interior of the apartments was adorned with incomparable magnificence, and their halls and porticoes resembled veritable museums, where the masterpieces of all ages had been gathered. Pliny already said that in his time the works of the most distinguished artists of Greece were seen there, and the Emperors who followed, especially Hadrian, that refined connoisseur and passionate lover of the Arts, must have enormously enriched the collection. In order that nothing might be wanting, rare and precious books had also been collected in abundance. The two libraries, Latin and Greek, of the portico of Apollo and of the house of Tiberius, were world-renowned.

Let us finally add that the situation of the imperial palaces was in keeping with their beauty. Cicero says that the Palatine was the finest spot in Rome. Thence one commanded the whole town, and the eye embraced nearly all the famous monuments with which the Republic and the Empire had adorned it. "What more noble abode," says Claudian, " could the world's masters choose? On this hill power is more majestic, and seems more conscious of its might. Here the palaces of the monarchs, raising their proud heads above the Forum, see at their feet the temples of the gods ranged round them in a circle, like outposts to protect them. Sublime sight! Thence the eye perceives above the altars of thundering Jove the giants hanging on the

Tarpeian Rock, the chiselled gold of the doors of the Capitol, and on the pinnacles of the temples, which on all sides usurp the realms of the air, those statues which seem to move in the clouds. Further on are rostral columns covered with the brass of ships, those edifices built on the top of the highest hills, audacious works which the hand of man adds to the work of nature, and those innumerable triumphal arches, fraught with the spoils of nations. Everywhere the splendour of gold smites on the dazzled sight, and by its ceaseless gleaming tires the trembling eyeballs."[1] All these riches have passed away. Nothing but their foundation remains of those marble palaces, from whose height the poet viewed the gold-encrusted buildings of the Forum. To-day they are only ruins, from which the eye looks forth on other ruins. But if we feel it difficult to imagine what they must have been when entire, let us remember that those who visited them in the last years of the Western Empire were unable to believe that magnificence could go further, and that they seemed to them the ideal of a regal habitation. From the third century, the word *palace*, derived from the name of the Palatine, designates in Latin and Greek the abode of a monarch. Thence it passed into modern language, like that of Cæsar, which the barbarians piously gathered up at the very moment when they were destroying the Empire, in order to make it the finest title that could be given to supreme power.

---

[1] Claudian, in *sext. cons. Honorii*, 35.

CHAPTER III.

THE CATACOMBS.

THE discoveries made for thirty years past in the Catacombs[1] present two remarkable peculiarities. First of all, they are the work of a single man, and it may be said that Signor J. B. Rossi shares their glory with no one; and then they have this characteristic — that chance has nothing to do with them, that they are the reward of a confident science which proceeds with order, and in accordance with fixed and certain rules. Signor Rossi never works at random. He knows what he is doing and whither he is going, and always announces in advance what he is about to find. Nothing shows better than the brilliant success of his excavations the advantage of a good method to works of this kind.

The Catacombs, which had not been visited since the ninth century, and whose memory was almost lost,[2]

---

[1] I call these monuments *Catacombs*, merely in order to conform to custom. In fact, only those of St Sebastian are so termed. The only name suited to them is that of cemeteries, and it is seen from a passage of Eusebius (*Hist. eccles.*, VII. 11) that the name catacomb was reserved for Christian burial-places.

[2] Signor Rossi found, however, in the Catacombs of St Calixtus, the names of Pomporius Læto and other scholars of the fifteenth century, who style themselves *antiquitatis perscrutatores et amatores*.

were found again by chance in 1578. Some years later, Bosio, an illustrious *savant*, undertook to study them, and having a clear-sighted, accurate mind, he at once found the means to render this study fruitful. He began by making himself familiar with the whole of Christian antiquity, so that, thanks to his immense reading, he was enabled to approach the Catacombs furnished with documents that would enable him to understand them. He purposed to explore them one after the other, to follow each regularly through the labyrinth of its galleries, and endeavour to find its name again and remake its history. Such a work demanded infinite erudition, a profound knowledge of the ecclesiastical authors, and marvellous efforts of sagacity. Bosio was doubtless equal to it; his successors seemed terrified at the task, and abandoned it. They neglected more and more to busy themselves with the Catacombs themselves, in order to concentrate their attention on the monuments that were discovered there. In the visits which they made to them, they copied the inscriptions and the paintings, without indicating the spot where they had found them, taking away all that could be taken, and placing it in some museum, and there the work of art, isolated from its surroundings and detached from the walls for which it had been made, lost its character and its importance. These curiosities of detail, which should only be accessory,

---

Being much suspected of recurring to paganism, and watched by the Popes, they hid their meetings in the Christian cemeteries, where they were sure of not being followed. Is it not singular that after having sheltered the first Christian assemblies, the Catacombs should serve as an asylum for the pagans of the Renaissance?

caused the specific study of the cemeteries, which is the essential, to be neglected; and the mine whence so many precious objects came was thus overlooked for the sake of the riches that were drawn from it. This, however, was the manner in which all ancient monuments were explored, and which proved so fatal to them.

Signor Rossi resolutely changed this method. He dared to say that for two hundred years past the right road had been abandoned, that all his forerunners had been mistaken, and that it was necessary to get upon Bosio's track again, and resume the work where he had left it. He rightly maintained that, in order to draw more profit from the venerable remains of ancient Christianity, they must not be separated from the study of the spots where they were placed, and that if they deserve to be studied on account of the memories they recall, still more is it necessary to be well acquainted with the Catacombs themselves, which are the most astonishing work of budding Christianity. This is why he proposed, like Bosio, to study the different Christian cemeteries successively; to design their plan; to search out the primitive extent of each, with the additions it has received; to determine, as far as possible, the period when each gallery was hollowed out, which at the same time helps to determine the age of the monuments it contains; and, in a word, to discover the history and settle the topography of this immense subterranean city, as has been done with so much success for the one that was built above it.

Such are Signor Rossi's aim and the method he has

professed to follow. We are about to see the results of his labours.[1]

I.

THE IMPORTANCE WHICH CHRISTIANS ATTACHED TO SEPULTURE—THE CATACOMBS THEIR WORK, AND NOT OLD ABANDONED QUARRIES—HOW THEY WERE INDUCED TO HOLLOW THEM OUT—HYPOGEA OF DIFFERENT RELIGIONS IN THE ROMAN CAMPAGNA — RULES ADOPTED BY THE CHURCH FOR BURIAL.

THE Catacombs are the place where the first Christians buried their dead. In the last century some scholars held that they might have served as a common cemetery for the poor of all religions, but this is an opinion which it is now no longer possible to maintain. For thirty-five years past the works have been pushed on with vigour, and thousands of tombs have been discovered, yet in not a single instance was a pagan tomb met with. It may therefore be fearlessly affirmed that they were reserved for Christians alone.

The Christians attached great importance to sepulture. The body being destined to come to life again, and share the soul's immortality, they thought that it should be taken care of after death, and given an

---

[1] I am about to expound them rapidly, according to Signor Rossi's great work (*La Roma sotterranea cristiana*, 3 vols., 1864–1878). Among the books in which Signor Rossi's researches have been presented to Frenchman, I will cite the *Nouvelles études sur les catacombes*, by M. Desbassyns de Richemont, and above all the translation of Messrs Northcote and Brownlow's book, published by M. Alard under the title *Rome souterraine*, Paris, Didier, 1872.

honourable asylum while waiting for the great awakening. "Soon," said Prudentius in his burial hymn, "soon the time will come when heat shall revive these bones, when blood shall gush anew in these veins, when life shall resume this abode which it has left. These bodies, long inert, which lay in the dust of tombs, shall spring upward once again to join their former souls." And he added in admirable lines: "Earth, receive and keep in thy maternal breast this mortal spoil which we confide to thee: it was the dwelling of a soul created by the author of all things; 'twas here a spirit lived, quickened by the wisdom of Christ. Cover this body which we place within thy breast. One day He who created it and fashioned it with His hands, will ask thee for His work again." No one being excluded from this hope, the Christians took equal care for the interment of all the faithful. They would have been horrified to imitate the pagans, and fling the bodies of poor people into common graves (*puticuli*) to rot. We see that it was forbidden among them to place two bodies one above the other. Each was to have its own place wherein to repose alone until the last day. We know from Tertullian that a priest assisted at burials;[1] the religion consecrated tombs. At the time of the persecution of Decius, the Roman clergy, writing to their brethren at Carthage, reminded them that there is no more important duty than to give sepulture to martyrs and other Christians.[2] The treasury of the Church was spent in helping the poor to live, and in properly burying them. Finally, St Ambrose agreed

---

[1] Tertullian, *De anima*, 51. [2] St Cyprian, *Epp.*, 8.

that it was rightful to break, cause to be melted, and sell, the sacred vessels for the interment of the faithful.[1] These texts explain the construction of the Catacombs. Knowing the respect shown by the first Christians towards their dead, we are less astonished at the gigantic works which they undertook for their burial.

But do these works in truth belong to them? Are the Catacombs entirely the work of the Christians, or were they merely appropriated to their use? This question has given rise to great discussions. In the last century there was no lack of incredulous persons who denied the reality of Bosio's discoveries. When told that the first believers themselves dug out their cemeteries, they asked who had furnished a small and poor community with the means needed for the piercing of this tremendous number of underground galleries; what could have been done with the earth drawn from them; and how the members of a proscribed religion could have had the audacity thus to dig out the ground at the gates of Rome, and before the eyes of their persecutors? These objections seemed to the majority of scholars unanswerable, and troubled even the most intrepid defenders of the Catacombs. So, in reply, they thought it well to suppose them ancient quarries, whence the Romans had for a long time dug *pozzolana*. The Christians had found them deserted, and, in order to convert them into their cemeteries, they only had to hollow out horizontal niches for the reception of the dead. The existence of these quarries was not a hypothesis; it is attested by ancient writers.

---

[1] St Ambrose, *De off*. II. 142.

Cicero speaks of a man who was in his time murdered in them,[1] and Suetonius relates that when they tried to persuade Nero to take refuge there, he declared that he would not bury himself alive.[2] Being a little-frequented place, where people wishing to hide themselves could find an asylum, they suited the Christians for the purpose of celebrating their mysteries and burying their dead. Bottari bids us remark that they might easily have been known to them. Their religion was first spread among poor people and slaves—that is to say, among the class employed in digging the quarries. These were so many guides, who could lead their brethren through the turnings of the deserted galleries. This opinion, therefore, appeared perfectly probable. It had the advantage, too, of silencing the incredulous, so it was religiously accepted by every one for two centuries, and down to our time was received without dispute. However, it does not hold good before an attentive examination of the Catacombs. Père Marchi began to undermine it, and Signor Rossi toppled it down. He has no trouble in showing that chambers 3 or 4 mètres square, and galleries 1 mètre, at most, in width, would have been scarcely convenient for the extraction and transportation of *pozzolana*. Ancient Roman quarries exist, whose destination is not doubtful, and their appearance is very different from that of the Catacombs. The passages are wider, and the outlets multiplied. Everything about them appears more suited to the necessities of an industrial exploitation. Furthermore, Signor

---

[1] Cic., *Pro. Cluentio*, 14.   [2] Suet., *Nero*, 48.

Michaele Rossi[1] on carefully studying the nature of the ground in which the cemeteries of Rome are dug, remarked that the workers systematically avoid the banks of friable *pozzolana*, preferring to bore where the stone is spongier and harder, and he declares roundly that never could materials suited for building have been dug from the Catacombs. This reason is decisive, and clears up the last possible doubts that could have existed. This does not mean that the Christians did not occasionally appropriate to their use some of those abandoned quarries called *arenariæ*. History says that they did so, and the researches made of late years prove it. I will say later, on what occasion and by what motives they were led to do so in very exceptional cases. In fine, in the twenty-five or thirty cemeteries hitherto visited, only five of these ancient quarries have hitherto been recognised, and there are probably not many more. All the rest were made by the hands of the Christians. Drawings of diggers at work are often seen in the Catacombs. They are represented, pick in hand, attacking the overhanging rock. The attitude given to them, represents the manner in which they proceeded. They advanced boldly, making themselves a way through the strata of granular tufa, of which the soil of the Roman Cam-

---

[1] Signor Michaele Rossi is the brother of Signor J. B. Rossi. He had received the education of a lawyer, but became a geometrician from inclination. The desire to help his brother, who needed an associate to study the soil and design the plan of the galleries, developed in him a vocation which he did not know himself to possess. He soon made himself a name in this new science, and has even invented an ingenious machine to shorten the work of plan-raising, which took a medal at the London Exhibition.

pagna is composed; they dug the rock before them, sustained by their faith, "living in the entrails of the earth, like the monk in his cell," and these interminable galleries, said to contain 6,000,000 tombs, are entirely their work.

Where did the first Christians learn this mode of sepulture, which required such terrible labour? It has long since been answered that they learnt it from the Jews. It should have been added that in this the Jews only followed the custom of most of the peoples of the East. There was no other mode of interment in Syria. Everywhere where the Syrians penetrated —in Malta, in Sicily, in Sardinia—similar burial-places are found. M. Beulé has confirmed the existence of Catacombs at Carthage; M. Renan saw them in Phœnicia: Asia Minor, Cyrenaica, and the Chersonese contain a great number, and they occur even among the Etruscans, to whom an Oriental origin is sometimes attributed. Lastly, they are found every day at Rome, and this must not surprise us. At the end of the Republic and the beginning of the Empire, Rome was, in a manner, invaded by the nations of the East. They brought to this great, tolerant, distracted town their beliefs and their customs. They were allowed to pray to their gods after their own fashion, and to bury their dead as they chose. Not only were they unmolested, but they were allowed to preach their doctrines, and did not fail to do so. I do not believe that any town, even Alexandria under the Ptolemies, offered to the world a more curious and animated spectacle than Rome at the beginning of the Empire. Not only was it the industrial and political capital of the world, it was also the spot where all the philosophies and all the religions of the world met. In the midst of enormous business activity, there reigned an activity of mind more remarkable yet. The weakening of ancient beliefs left the field open to new opinions, so that the Christians profited by the circumstance to agitate and spread, and made proselytes everywhere. The religions of the East especially attracted men's souls by the strangeness of their rites and the mysterious turn of their doctrines. Some quite yielded themselves up to them; the greater number, without being entirely permeated by their spirit, at least imitated their commonest practices. It is thus that many Romans took to burying their dead in the manner of the Orientals. From the time of the Antonines, the custom of burning the dead became less and less frequent, and at the time of Macrobius it scarcely existed at all.[1] So the pagans at an early date possessed *hypogea*, like those of the nations of the East. I imagine that from the end of the second century the Roman Campagna must have been dug in every direction. The Jews, the Phœnicians, the worshippers of Mithra and Sabazius, and, above all, the Christians, who were becoming so numerous, and sometimes the pagans, hollowed out the ground for their sepulchres. There was in these various religions a kind of interior and subterranean activity, corresponding with that outside. These sepulchre diggers sought to avoid each other,[2] but they did not

---

[1] Macrobius, *Sat.*, VIII. 7.

[2] Signor Rossi shows that more than once the Christian galleries have abruptly turned aside in order not to touch some *hypogeum* of another cult.

always succeed. In the heart of the Catacombs a cave is found where a priest of Sabazius and some of his disciples rest. The Christian workmen doubtless came upon it suddenly, and it now communicates freely with the tombs of the martyrs. The number of crypts that were then dug are incalculable. Fresh ones are discovered each day. Pagan *hypogea* are no longer rare. The names of more than forty Christian cemeteries are known. We are acquainted with two Jewish Catacombs—that of Trastevere, which is anterior to Christianity, and that on the Via Appia, and it is to be hoped that more will be found that will teach us what we should so much like to know—the constitution and government of the synagogues at Rome. Perhaps, too, we may come upon those of the dissenting sects of Christianity. We know that they, too, had Catacombs, and that in order to give them some authority, they went and stole the bodies of the most respected martyrs from the orthodox Catacombs, and placed them with themselves. What light will not these discoveries throw upon the religious history of those times, if they are always directed by men of honesty and science, like Signor Rossi?

Among all those burial-places, which so much resemble each other, the Christian cemeteries may be recognised by two signs. In the first place, they are much more extensive than the others. Nowhere have such a development of galleries and such an accumulation of tombs been found, and never did any religion or any nation seem to feel so strongly as the Christians the need to group together and unite in death. Then the niches containing the bodies are open in the Jewish crypts and closed in the Christian Catacombs. This difference is connected with the habit which the Christians had of assiduously visiting the tombs of the martyrs. With the Jews, who only opened the sepulchre when some one was to be buried, it was not necessary to protect the body from the indiscreet curiosity of visitors; it was enough to roll a great stone to the entrance of the cave. It was different with the Christians, and their cemeteries being open to the faithful, their tombs of necessity had to be shut. In all else their Catacombs exactly resembled those of the Jews and other peoples of the East, and a first glance suffices to show that they learned this mode of burying their dead from them.

Yet it must not be thought that a fixed rule and constant custom existed in the primitive Church as regards burial. The only law accepted by all, was not to use for one's self or for one's relations the tombs of pagans, and not to admit pagans into the cemeteries where the Christians slept. "Let the dead bury their dead," harshly said St Hilary, and we know that in the time of Cyprian a forgetfulness of this law occasioned the deposition of a bishop. Beyond this, the faithful were free, and they used their liberty. So we sometimes see them use isolated tombs. The epitaph of two spouses has been found, who had a resting-place made for them in their garden (*in hortulis nostris secessimus*), and who do not appear to excuse themselves for doing so. Another gravestone contains a selfish inscription, a strange mixture of pagan habits with Christian terms, by which the possessor of the tomb cites before the judgment of the Lord whoever shall try to introduce another body into the grave he occupies or the grounds surrounding it. He wishes to have them all for himself alone. Yet the Christians were usually imbued with other feelings. As I said just now, they felt the need of resting together. They desired to be united in death, as they had endeavoured to be united in life. From the first days they grouped themselves instinctively round the bishops and the martyrs, and in the whole of Christendom those collections of tombs were soon formed, which received the name of places of repose or sleep (*accubitorium*, χοιμητήριον).

Only, according to the country, these cemeteries were situated in the open air or hidden under ground. At Rome, subterranean burial was preferred. Was this because the Christians were there more in the sight of the governing powers and more feared their supervision? More probably it was the order to remain faithful to the traditions of the newly-born Church, which, on leaving the Jewish community, had retained this of its customs. It was, above all, in order to imitate the tomb of Christ, whose life and death were the example of Christians. There can be no doubt that the sepulchre of Joseph of Arimathea, "which had not been used, and which he had caused to be cut in the rock," with its horizontal niche surmounted, for sole ornament, by an arched roof,[1] served as a model for

---

[1] These niches hollowed out in the wall are called *loculi*. The arched roofs surmounting them have received the name of *arcosolia*. These are not found on all the tombs, but only above those of the most important personages. More ample details respecting these

the first Christian tombs. We are, therefore, certain that the Catacombs were the work of the Christians, and that they were dug by them and for them. It was necessary to be sure of it before beginning our study. This point established, we can enter and go through them. Only let us be careful to put ourselves under the guidance of Signor Rossi, for he is the best guide we can choose, if we would visit them with profit.

II.

FIRST IMPRESSIONS PRODUCED BY A VISIT TO THE CATACOMBS—THE IMMENSITY OF THE CITIES OF THE DEAD, AND CONSEQUENCES TO BE DRAWN FROM IT— RAPID DIFFUSION OF CHRISTIANITY — RELIGION SEPARATES ITSELF FROM THE FAMILY AND THE COUNTRY — THE CATACOMBS THE MOST ANCIENT MONUMENT OF CHRISTIANITY AT ROME—MEMENTOES OF THE TIMES OF PERSECUTION CONTAINED IN THEM —MEMENTOES OF THE DAYS OF TRIUMPH.

A VISIT to the Catacombs, especially if prolonged for several hours, may possibly cause more surprise than pleasure to people unprepared for it by some preliminary

---

words will be found in Abbé Martigny's *Dictionnaire des Antiquités chrétiennes*. I profit by the occasion to recommend this excellent work, which is indispensable to all who would study the principles of Christian archæology, and useful to people of the world for the understanding of many words that are read and repeated without being more than half understood. On using it, they will be very grateful to the modest and distinguished man who has known how to put together so much solid knowledge in so convenient a form.

study. It will, perhaps, have but small effect on those to whom the history of the first years of Christianity is but little known. In any case it would lose much of its interest if one were not requested at every turn to remark certain particulars which of themselves scarce draw the attention, but which are, nevertheless, of great importance. At first all looks alike, and nothing seems particularly noticeable. We pass along narrow underground galleries, where it is difficult to walk two abreast, and we skirt walls pierced with parallel niches, very like great drawers placed one above the other, which were used for burial. When a body had been placed in one of them, the opening was closed by slabs of marble or bricks, on which the names of the deceased were inscribed. Almost all these bricks have been removed, and at the bottom of the niches the little heap of dust which, after fifteen centuries a decomposed body forms, is easily seen. From time to time we meet on our way more roomy and more ornate chambers for the dead of note. They usually contain paintings, nearly effaced, of which it is very difficult to seize even a few details by the doubtful light of the *cerini*, and which, when looked at rather hurriedly, appear to resemble each other very much. The galleries cut each other at right angles, forming a tangled labyrinth in which it is impossible to find one's way. When we have traversed one storey, staircases lead to a lower one, where the same spectacle is repeated, only the darkness seems to have doubled, breathing becomes more painful, and the heart is more and more oppressed as we plunge deeper into the earth and leave air and light farther and farther behind.

Having overcome this first impression, we begin to reason and reflect. First of all, it is difficult when the visit is prolonged not to be greatly struck with the great immensity of these cities of the dead. These superposed stories, these galleries added unceasingly to each other, these graves crowded more and more thickly along the walls, are a startling image of the rapidity with which Christianity spread in Rome. The first who buried their dead in the Catacombs do not seem to have expected such rapid progress. They were content to hollow out a few galleries close to the surface, and encumber them with huge sarcophagi placed against the wall. But soon, the ranks of the faithful augmenting, the number of the dead grew so considerable that it became impossible to take things so easily. It has often been asked whether there is not a great deal of exaggeration in those passages where the Fathers of the Church describe the marvellous development of Christianity to us, and show it to us from the end of the second century, filling "the cities, the islands, the castles, the camps, the tribes, the palaces, the Senate, the Forum, and only leaving to the pagans their temples." It must be owned that the indefinite increase of the cemeteries, with the necessity of constantly adding new galleries to the old, and of crowding the tombs one against the other, seems to prove them entirely in the right.

The immense extent of the Catacombs ere long suggests another reflection, not devoid of importance. The pagan cemeteries, to which we cannot help comparing them, were much less vast; they generally held but a single family. The largest are those containing the freedmen of the same master, the members of the same college, or poor people who had joined together, in order to build themselves a common tomb at less expense. It was another bond that united those who chose to sleep together in the Catacombs. Their country, their birth, their fortune, were often very diverse; they belonged to different families; they did not pursue the same calling; and perhaps some of them never met during their lifetime. Their only bond of union was religion, but this bond became so powerful that it replaced all others. We have just seen that the Church did not impose common burial upon the faithful as a duty, and that there were some among the first Christians who caused private tombs to be constructed upon their domains, to which they did not admit their fellows;[1] but those must have been very rare, and almost all chose to be buried with their brethren. When we reflect on it, we will see that this was a serious innovation, and a new manner of considering religion. Among almost all ancient nations, religion did not separate itself from the family and the country. Christianity first divided that which all Antiquity had united. Henceforth domestic or national gods ceased to be worshipped, and religion had its own independent existence outside the family and the State, and above them. Many of those who are buried in

---

[1] A few family tombs have also been found in the Catacombs, but they could not have been numerous. The earth dug from new galleries was generally used to fill up the old ones in which there was no more room. Thus it became impossible for a family to keep a tomb beyond one or two generations.

the Catacombs doubtless possessed domestic tombs elsewhere; others might have been buried among people of their own condition with whom they had passed their lives, but all of them chose to rest in one of the great Christian cemeteries. They voluntarily renounced that neighbourhood of relations and friends, which until then had been regarded as one of the great consolations of death. They took their assigned place among strangers, who often came from the most distant countries, and to whom nothing attached them but their belief. Slaves, freedmen, and citizens; Greeks, Romans, and barbarians, forgot all these diversities of fortune and origin, and remembered only their common religion. Nothing was more opposed to ancient society than the separation which was then effected between the family or the State and religion. It is the work of Christianity, and it is in the Catacombs that it is most evidently manifested.

Such are the reflections that first occur to the mind, even when we are content just to go rapidly through these long galleries. If we have time to look at them more closely, our interest and curiosity increase. Let us reflect that the Catacombs are the most ancient monument of Christianity at Rome. The others only date from the fourth century—that is to say, from a time when the dogma was already fixed; when the new religion had gained power and had found a language wherewith to express its tenets.

In the Catacombs the history of primitive Christianity is almost complete, and in going over it we may follow all the vicissitudes of its agitated existence. These galleries which lead out freely upon the great public ways, these openings destined to give a little air and light to the *hypogea*, are of a time when the Christians were tolerated, and confided in the protection of authority. These dark entrances, on the contrary, and these tortuous ways, recall the period of the persecutions. It is then that those little chapels were constructed, where the faithful assembled when they could no longer worship in open day. They are usually composed of two chambers, crossed by the gallery of the Catacombs, so that they are separated from each other, yet at the same time near enough for it to be possible to follow the sacred services in both of them. They were destined for the two sexes who, in the primitive Church, were never united. At the end of one of the chambers is the stone seat on which the priest took his place to celebrate the holy mysteries and talk to the assembly. There it is that words of exhortation, such as we find in the works of the fathers, must often have been pronounced, to kindle those present and give them courage to brave death for their faith. Here the letters were read, addressed by one church to another, to communicate their fears and hopes, and spur each other on to endure and express their beliefs. Not one of them recalls the period of gropings and struggles; not one of them has preserved relics of the heroic age of the Church. Furthermore, they have been too often restored and remade; they have assumed too modern an air. What is there remaining about the basilicas of Constantine really antique? What trouble do we not have to picture to ourselves what St Laurentius, St Praxedes, or St Agnes were like when just built? The Catacombs are better preserved. They have the good fortune to have been nearly forgotten and lost down to Bosio's time. If, since then, they have sometimes happened to be devastated by greedy amateurs or clumsy explorers, it has, at least, occurred to no one to rebuild, under pretence of repairing them. They are the most venerable remains, the most authentic witness of the first centuries of Christianity, and there is no monument in Rome that better puts us in presence of those primitive times which are so little known to us and with which we so much desire to become acquainted.

Everything now becomes interesting, and the least details now assume importance. We must raise with care these bricks which have been loosened from the tombs, and which travellers tread under foot; for they may bear the mark of those who made them, and help to fix the date of the galleries. From time to time, our attention is called to a little niche in the sombre walls we are passing, or to a console jutting out; here the clay-lamp was placed to light visitors. How many times have friends or relations passed before it, to pray and weep by a cherished tomb! We pause a moment in those chambers, more roomy than the others, at the end of which we find a tomb disposed in the form of an altar. Signor Rossi tells us that they were used for family meetings. People gathered there on funeral anniversaries to implore the mercy of God for the departed: "To read together the holy books, and to sing hymns in honour of the dead who sleep in the Lord." It is easy to imagine the effect which such ceremonies must have produced upon pious souls. In the midst of this solemn silence, between these walls lined with corpses, they seemed to live quite in the company of those they had lost. The emotion which seized them brought home more clearly that oneness of the dead with the living which paganism had recognised, and the Church made one of its dogmas. They felt so full of all these dear memories that it required no effort to believe that death cannot break the bonds which bind man to man, and that they continue to render each other mutual services beyond life—those who are no more profiting by the prayers of the Church, or, if they enjoy celestial beatitude, helping those who still live by their intercession.[1] This is the sentiment expressed by the pious exclamations which visitors have traced on the wall with the point of a knife, and which Signor Rossi, not without great trouble, has succeeded in copying and making out. It was here, too, after those great persecutions which increased the number of the martyrs, that they took comfort together, encouraged each other to continue, and celebrated the memory of the dead, both glorifying them and glorying in them for the example they had given to the community of the faithful. "Happy our Church! The Lord protects and honours it. It was, till now, shining white by the good works of our brothers. He vouchsafes it the glory of being reddened with the blood of the martyrs; neither lilies nor roses are wanting in its crown!"[2] The epoch of the persecutions

---

[1] These expressions are borrowed from one of the most ancient rituals of the Roman Church, cited by Signor Rossi: *Defunctorum fidelium animæ quae beatitudine gaudent nobis opitulentur ; quae consolatione indigent Ecclesiæ precibus absolvantur.*

[2] St Cyprian, *Epist.*, 10.

seems to have remained more vividly impressed upon the Christian cemeteries than all the others, and Signor Rossi shows us traces of it everywhere. He points out how the old staircases were then demolished and the great galleries filled up, in order to shelter the tombs of the martyrs from profanation. New roads were hurriedly made, leading to those abandoned sand-pits (*arenariæ*) of which I spoke just now. That way they could go in and out without arousing suspicion; and they endeavoured to make even these secret issues impracticable for strangers and invaders. Signor Rossi found in the gallery of Callistus a staircase the steps of which are abruptly interrupted. It was only possible to proceed thence into the interior galleries by means of a ladder, placed by an accomplice at a given signal, and which he withdrew when all the faithful had entered. But these minute precautions did not always suffice to save the Christians. We know that there were spies and traitors among them who warned the police. "You know the days of our meetings," said Tertullian to the magistrates, "you have your eye upon us even in our most secret meetings; so you often come to surprise and overwhelm us!" The emperor's soldiers more than once penetrated into the Catacombs, interrupting the ceremonies, and striking down without pity all whom they could seize. Inscriptions, of which a few fragments have reached us, preserve the memory of these sanguinary executions. Perhaps one day that chamber will be found, where some unfortunate Christians, surprised in the act of celebrating their worship on the tomb of a martyr, were walled up and left to die of hunger. Pope Damascus, in repairing the Christian cemeteries, wished that the spot that witnessed this terrible scene should be respected. He contented himself with opening in the wall a broad window, whence the faithful could see the bodies stretched upon the ground, just as death had smitten them.

Beside these witnesses of proscription and mourning, the Catacombs retain traces of those days of triumph. Everywhere the remains of great works are seen, which were undertaken after peace came to the Church, in order to consolidate or embellish them. After Constantine, the practice of burying the dead in them was gradually discontinued, and henceforth they were only a monument surrounded with veneration. Pilgrims came to visit them from all the countries of Christendom. All wished to see the sepulchres of famous martyrs; all desired to bear away some pious memento of their journey. There was even a queen who sent a priest on purpose to collect and bring away oil from the lamps which burnt before the tombs of the saints. The invasions of the barbarians interrupted this worship. Alaric, Vitiges, Ataulf in turn devastated the Roman Campagna. In order to shelter the holy relics from these ravages, they had reluctantly to take them from their tombs and bring them to Rome, where they were distributed among the different churches. Henceforth there was no longer any reason to visit the Catacombs, and down to the sixteenth century almost their very trace and memory were lost.

L

III.

THE INSCRIPTIONS AND PAINTINGS IN THE CATACOMBS—CHARACTER OF THE MOST ANCIENT INSCRIPTIONS—THE BIRTH OF CHRISTIAN ART—THE FIRST SUBJECTS TREATED BY THE ARTISTS OF THE CATACOMBS—IMITATION OF ANTIQUE TYPES—REPRODUCTION OF CHRISTIAN SUBJECTS—SYMBOLISM—ORIGIN OF HISTORICAL PAINTING—TO WHAT EXTENT THE CHRISTIAN ARTISTS ADHERED TO ANTIQUE ART.

It might be feared, at first sight, that history would not be able to derive much profit from these thousands of tombs, all resembling each other, and enclosing a whole nation of unknown dead. But these monuments are not so mute as they seem; epitaphs are found on almost all, while some are ornamented with bas-reliefs or frescoes. These inscriptions and paintings seem to lend them a voice. All mutilated and incomplete though they be, they teach us something of the life and feelings of those who sleep in the Catacombs.

The most ancient of the inscriptions are written in Greek. At the beginning of the third century this was still the official language of the Church. Latin only came later on. Among the epitaphs of the Popes, found by Signor Rossi in the cemetery of Calixtus, that of St Cornelius, who died in 252, is the only one in Latin. Greek appears only to have been abandoned little by little, and with regret. Some curious inscriptions enable us to note the change from one language to the other, and they show us the scruple felt at leaving that which the Church had used almost since its origin. In several of them the Latin words are written in Greek characters, and there are some in which the two tongues somewhat strangely mingle (*Julia Claudiane in pace et irene*). Only in the most recent galleries does Latin dominate almost exclusively.

The characteristics of the most ancient epitaphs are great shortness and simplicity. Christian epigraphy of the first times loved the garrulity of Greek inscriptions no more than it did the majestic solemnity of Roman inscriptions. It was content to write one of the names of the dead (we know that under the Empire it was a sort of distinction to bear many), and it added thereto some pious exclamations, all signifying much the same thing: "Peace with thee!" "Sleep in Christ!" "May thy soul rest with the Lord!" The time the deceased has lived and the date of his death are rarely mentioned; what are these earthly memories to him who has taken possession of eternity? While the pagans took great care to inscribe on tombs the dignities filled by the deceased, and the rank which he held in life, they are never mentioned among the Christians. "With us," said Lactantius, "there is no difference between the poor and the rich, the slave and the free man. We call ourselves brothers, because we believe ourselves to be all equal."[1] Do what we may, equality always suffers

---

[1] The Christians did not act thus in obedience to an express and imposed rule, but from a sort of common and spontaneous feeling. This is proved by the circumstance that in the crypt of Lucina, the most ancient part of the cemetery of Calixtus, mention is made of a freedman; and although ecclesiastical dignities are usually no more recorded than others, we see that there were three priests, and

during life; the brothers were at least resolved to preserve it in death. Their heroic humility is attended with some inconvenience so far as we are concerned, and the silence to which they condemned themselves deprives us of a mass of curious information. We learn much, however, from what they choose to tell us. Their epitaphs show us that certain opinions, which are sometimes thought new, existed in the Christian community from the end of the third century. They believed, for instance, in the efficacy of the prayers of the living for the dead. The exclamations I have just quoted are more than wishes — they contain requests made to God, which are supposed to be listened to. The intercession of the saints for those who pray to them was believed in. The faithful, who visited the tomb of a martyr with such fervour, indeed thought that he took an interest in their weal, and would help them to attain it. In one of the inscriptions gathered by Signor Rossi, a young maiden is addressed who has just died, and who is believed to be a saint, and they tell her: "Invoke God for Phœbe and for her husband (*pete pro Phœbe et pro virginio ejus.*"[1])

Later on, this primitive simplicity of the inscriptions

---

that one of them was at the same time a priest and a physician. It was not therefore absolutely forbidden to preserve the memory of social distinctions in Christians epitaphs, and the prevailing abstention from so doing was voluntary.

[1] By *virginis* was meant a man who had no other wife. It is not, as might be thought, quite a Christian term. The pagans used it. If they did not condemn second marriages so severely as certain rigid Christians did, they at least desired to show respect and approval towards those who had not made a bad use of the facility of divorce.

underwent a change. Regrets first appear, and it was difficult, indeed, for faith always to be strong enough to restrain them. Then a timid compliment to the dead was indulged in. Of a young girl it was said that she was an "innocent soul," or "a dove without bitterness;" while a man was called "very holy," or even "incomparable." The number of years he had lived was noted, and the exact date of his burial, or, as they said, his *deposition*. At length these details were reproduced in like manner on all tombs. The style of Christian inscriptions was then fixed, or, in other words, formula and convention slipped into a place where only an impulse of the heart should ever have been found. I well understand that there would be some to whom this progress is not altogether pleasing. In presence of these inscriptions of the fourth century, so ordered and so regular, it is difficult not to regret the time when sorrow and faith were less disciplined, and when each expressed his regrets and his hopes as he felt them, without being careful to follow custom, and weep like every one else.

The paintings are still more important than the inscriptions; for they enable us to go back to the beginnings of Christian art. As it arose from the worship of the dead, its first attempts were naturally made in the Catacombs. The Christians were very anxious by every means to honour the sepulture of those they had lost, especially when they had died victims to some persecution. Doubtless, sculpture and painting must have seemed to them profaned by the every-day use which the pagans made of them; yet they did not hesitate to adopt them for their cemeteries.

They perhaps thought that by employing them to embellish the last abode of their brothers, they purified them.

The first artists who were called upon to decorate the Christian tombs with frescoes or bas-reliefs, were in all probability very much embarrassed. What subjects should they represent? For an art at its very beginning, the question was a grave one. As the Christian sect was proscribed, and their doctrine had to remain secret, they naturally at first used certain signs of which they alone knew the true meaning, in order to recognise each other. This was done in the pagan mysteries. We know that objects were distributed among the initiated which they were to keep, and which were to remind them of what they had seen during the ceremonies of initiation.[1] It was the same with the first Christians. Clement of Alexandria reports that they had engraved upon their rings the image of the dove, of the fish, of the ship with sails spread, of the harp, of the anchor, etc.[2] These were symbols which recalled to them the most secret truths of their religion. Almost all these symbols are found again in the Catacombs; but they are not alone there. Signs so dark and vague could not suffice the faithful, and the sculptors and painters whom they employed, and who were for the most part deserters from paganism, had to endeavour to represent their new beliefs in a manner more direct, more clear, and more strictly artistic. But everything here had to be created. Since the Jews offered no model in this respect, they were of necessity

---

[1] Apuleius, *De magia*, 55. [2] Clement of Alex., *Pædag.*, III. 11.

obliged to apply elsewhere, and take art where it was to be found,—that is to say, in the pagan schools. They did so without scruple, so long as only those simple ornaments were in question which had no real meaning, and were found everywhere. Tertullian himself, the severe doctor, allowed them to do this.[1] In order to adorn the walls and roofs of their funeral chambers, they copied the graceful decorations commonly used in the houses of the pagans. Ceilings of this kind are numerous enough in the Catacombs, and there are some in the cemetery of Calixtus which may be placed among the most pleasing that Antiquity has left us.[2] As at Pompeii, we see charming arabesques, birds, and flowers, and even those winged genii which seem to fly in space. Is it not strange that these marvels of grace and elegance, in which there breathes all the smiling art of Greece, should be found amid the dark galleries of an underground cemetery? We must believe that the details and emblems of this decorative painting, by dint of being used in over-profusion, had lost all meaning for the mind. It remained only a pleasure for the eyes, and no one was scandalised or even surprised at seeing them reproduced above the tomb of a believer. But the Christian artists dared more. It being difficult for them suddenly to invent an original expression for their beliefs, they imitated some of the purest types of classic art, whenever they could be allegorically applied to the new religion. This imitation is seen already in "The Good Shepherd," which seems to have been inspired, at least as to its first idea

---

[1] Tert., *Adv. Marc.*, 11, 29.   [2] Rossi, *Roma sott.*, X. 18.

and general composition, by some ancient paintings.[1] It is more evident yet in those fine frescoes where the Saviour is shown with the attributes of Orpheus. The singer of Thrace, drawing the beasts and the rocks by the sound of his lyre, might seem an image of Him whose word conquered the most barbarous nations and the lowest classes among civilized people. Three reproductions of this subject are known in the Catacombs. The sculptors do the same as the painters, and even go farther still. The painters worked in the Catacombs themselves, far from the indiscreet and the unbelieving, and their frescoes were imagined and executed in this silent city of the dead, where everything invited the artist to yield himself up to the ardour of his faith. The sarcophagi were worked in the studios, where all could see them, and this necessitated prudence. It is even probable that most often, when the Christians wanted a tomb of stone or marble, they took it ready made from the merchant, choosing that in which the figures were least shocking to their opinions. It is thus that some are found in the cemetery of Calixtus representing the adventure of Ulysses with the Sirens, and the poetic story of Psyche and Cupid.[2]

But Christian art was not long to live by borrowing. A doctrine so young, so full of sap and life, which took possession of the entire soul and transformed it, must

---

[1] Rossi, *Roma sott.*, I. 347 : *In quanto però alla composizione artistica del gruppo, nulla osta a credere che i primi pittori cristiani abbiano potuto imitare, per quanto al loro scopo si confaceva, qualche bel tipo d'un simile gruppo di antico e classico stile.*

[2] As a matter of fact, the figures of this last sarcophagus had been covered with lime. But there are others concerning which the same scruples had not existed. See on this subject, Collignon, *Essai sur les monuments relatifs au mythe du Psyche*, p. 436, etc.

soon come to express itself in a manner of its own. We have already remarked that even when it borrows types not belonging to it, it fashions them in its own way, and seeks to appropriate them. The Orpheus of the cemetery of Calixtus, instead of drawing the beasts and the trees to him, as the fable relates, and as he is represented at Pompeii, has only two sheep at his feet, which seem to listen to his songs. We see that he is in process of confusion with the Good Pastor. Soon the artists dared to draw their inspiration directly from their belief, and to represent events taken from the holy books. From the Old Testament there was "The Sacrifice of Isaac," "The Passage of the Red Sea," the history of Jonah, of Daniel, of Susanna, of the Three Children in the Furnace; from the New Testament: "The Christ Child visited by the Magi," "The Cure of the Paralytic," "The Raising of Lazarus," and the "Multiplying of the Fishes." It has been remarked that they always abstain from recalling the painful circumstances of the Passion. Did they fear by representing Christ dying an infamous death, to scandalise the weak, to give scoffers a subject for ridicule, or to fail in respect towards their God? What is certain is, that they never represented the scenes which passed between the judgment of Pilate and the Resurrection. It is not uninteresting to remark that, on the contrary, the artists of the Middle Ages delighted to treat those subjects which their predecessors so carefully avoided; that they abounded in representations of the Flagellation and of the Crucifixion, and that these spectacles, by touching the faithful to the heart, served to give a wonderful impulse to popular devotion.

Among the questions which present themselves to the mind when reviewing the work of Christian painters and sculptors in the Catacombs, there are two, especially, to which it does not seem easy to reply. These artists did not treat without distinction all the subjects furnished them by the holy books; they only took a certain number of them, which they reproduced continually. Why did they prefer the former to the latter, and what was the reason of their choice? They often unite different subjects in a way that appears quite arbitrary, and place, one after the other, scenes that do not seem to follow, and have no connection with each other. Did they act at random, or must we believe that for these strange groupings they had some motive which it were possible to guess? Usually everything is explained by symbolism, and symbolism must certainly have played a great part in the beginnings of Christian art. It is known that the doctors of the Church, above all in the East, often understood the Bible narratives in a figurative sense, and that they loved to see in them moral allegories or prophetic images of what was to come to pass under the new law. In doing so, they followed the example of Philo, who took much pains to give the Old Testament a philosophic meaning, and who professed to find in it the whole of Plato's doctrine. Philo himself imitated those pagan theologians who, desiring to be at the same time philosophers and devotees, and preserve their respect for ancient beliefs without too much humbling their reason, regarded the legends of mythology as symbols or figures, hiding beneath a rude envelope deep and useful truths. Christianity inherited all this work of exegesis, and may be said to have sometimes found the legacy rather burdensome. One of the causes of the fatigue we occasionally experience in reading the Fathers of the Church, is the effort they are continually making to find for everything figurative meanings; the mixture of subtle interpretations and sincere outbursts, of touching simplicity and refined pedantry, of *naïveté* and scholasticism, of youthfulness with senility, which at every moment reminds us that Christianity was a new religion, born in an old epoch, and that even in the best works of its greatest doctors it has often two ages at the same time.

Like contrasts are found again in the works of art of the first Christians. It is natural that these artists, who followed the taste of their period, should have often given a symbolic meaning to the scenes they represented in their paintings or their bas-reliefs. They seem even to have sometimes made a point of telling us this. A fresco in the Catacombs represents a sheep between two wolves. Below it we read the inscription: *Susanna, seniores.* So, by the two wolves and the sheep, the adventure of Susanna is figured. Noah stretching his arm towards the dove who brings him the desired branch, was an image of the Christian arrived at the term of his voyage, saved from the perils of the world, and about to attain Heaven. And the proof of this is, that Noah is occasionally replaced on the sarcophagi by the deceased, irrespective of age or sex, so that, instead of the venerable patriarch, one is much surprised to see quite a young man, or even a young woman, coming out of the ark.

It is then certain that among the paintings or bas-reliefs of the Catacombs, there must be many containing images or symbols, and that, for instance, in the figure of Jonah cast forth by the whale, the paralytic cured, and Lazarus brought to life, the faithful of the early ages formed allusions confirmatory of their hopes of immortality. What they then easily recognised we have now great pains in divining. However, some experts have tried to give us the key of these allegories.[1] In the cemetery of Calixtus two very ancient chambers were discovered, close to each other, which were built at the same time and decorated in the same spirit—perhaps by the same artists. They have represented a series of scenes drawn from the Old and New Testaments, believed to be entirely symbolical, and to contain in a consecutive and almost dogmatic form the most sacred doctrines of the Christians. Signor Rossi undertakes to give the sense of all these symbols, whether by comparing the two chambers with each other or by invoking the authority of the Fathers of the Church.[2] He shows that the sacred books are here interpreted after the manner of Origines and his disciples. Nothing is more remarkable than to see how strangely allegory and truth are mixed up. The rapid succession, and even confusion of the sense proper and the imaged meaning shows how accustomed everybody then was to this subtle exegesis, and how easily the doctor or the artist was followed in his expository fancies. This personage striking the rock is now Moses and now St

---

[1] These explanations have been carried much too far, and symbols and images discerned everywhere. M. Le Blant has shown the temerity of these attempts in his *Étude sur les sarcophages d'Arles* p. 15, etc.

[2] Rossi, *Roma sott.*, II. p. 331.

Peter;[1] and the flowing water is not only that destined to refresh the Hebrews in the desert, it is the source of grace and life, which, a little further on, we see a priest use to regenerate a young man by baptising him; it is also the immense sea of the world, into which the holy fisher of souls casts his nets. From one scene to another, and often in the same scenes, allegories follow, destroy, confuse, and replace each other. Here the fish represents the believer conquered to the faith; elsewhere it is Christ Himself who, on the three-legged table, beside the mystic bread, offers Himself as sustenance to His disciples. The vessel from which Jonah is flung into the sea has a cross on its mast; it is at the same time the Church, which a contemporary of St Calixtus compares to a ship beaten by the waves, yet never sunk. If Signor Rossi's manner of explaining these paintings is the right one, it may be concluded that Rome did not remain so great a stranger as is supposed to those works of ingenious interpretation, of which the learned Church of Alexandria became the centre, and which for us is summed up in the great name of Origines. But at Rome the movement soon stopped. The Roman mind could not have had much taste for these refined allegories and these bold subtleties in which the Greek genius delights. It rather prefers to take things in their historical and real sense than to lose itself in symbolic interpretations, into which there always enters a little imagination. A lover of light, of order, and of discipline, it always seeks to submit individual wills to the general sentiment. Thus it does

---

[1] This allegory is certain. The name of St Peter is sometimes written above the personage striking the rock of Horeb to make the water gush forth.

not reject the formula which throws all ideas into a uniform mould, and affords it the spectacle—preferred by it to all others—the semblance of unity. The day when it became dominant in the Church it changed its character and its destinies. Perhaps, if the influence of the Jews and the Greeks had been stronger, it would have made it a community, and sometimes an anarchy of souls in search of the truth, discussing passionately the means to discover it, and seeking it by different ways. But, thanks to the Roman spirit which took possession of the Church, it became above all things a government.

Art, like everything else, felt this influence, and seems to enter upon new ways in proportion as the Roman spirit gains the upper hand in the Church. Signor Rossi shows that in chambers of somewhat more recent date than those of which I have just spoken, the frescoes are still fine, but have no longer the same character. Allegories become more rare, and those we find are not treated with the same freedom and the same variety. The age of historical painting begins; its birth is seen, as it were, in the Catacombs. Signor Rossi discovered a very curious picture there, and one which seems to represent an almost contemporaneous event. Erect upon a *suggestum*, a personage of grave and threatening look, dressed in the *pretexta*, and wearing a crown, angrily addresses a young man placed in front of him. Behind them, a man also wearing a crown upon his head, and with his hand placed upon his chin, seems to retire in displeasure. Signor Rossi beholds in this picture a scene from the persecutions, and it is, according to him, the interrogation of a martyr. The examining magistrate, the emperor perhaps, is represented with his usual attributes. The Christian has indeed the bearing of a man confessing his faith: his features breathe gentleness and resolution, and the artist has given his eyes a strange brightness. He looks at no one; he does not appear to listen to what is said to him, and is evidently filled with other thoughts. As for the person who is retiring, he is doubtless a pagan priest who has not been able to induce the believer to sacrifice to the gods. This is probably the most ancient painting of a martyr that we possess. It is the beginning of a *genre* which, from the fourth and fifth centuries, was to become very much in vogue.[1]

The Catacombs, while acquainting us with the beginnings of Christian art, afford us some particulars —the only ones we possess—touching the artists who decorated them : humble artists, who worked with such devotion in silence and darkness for the honour of their brothers much more than for the glory of their own names! Nothing has remained of them except their works; but the work enables us to guess the workman. Need we say that they were pious Christians and sincere believers? They must, indeed, have been so, to shut themselves up thus in these dark abodes, and paint pictures which no sunbeam was ever to illumine. But their piety did not induce them entirely to sacrifice their independence. They were not so much subdued by ecclesiastical influences as is thought, and it is not true to say, as has been asserted, that the Church held and guided their hands. The frequent mistakes they commit against the text of the sacred books shows that personal initiation, with all its errors and caprices, had

---

[1] Prudentius, *Perist.*, IX. and XI. 126.

some part in their works.[1] The resemblances remarked between them are less the effect of a command given or a direction received, than of a certain sterility of invention; the diversities, however slight they may be, prove that they did not work according to a unique and imposed model. Neither did they forget that they were artists as well as Christians. They did not think themselves free to withdraw from the eternal conditions of art, under pretext that they were working for a new religion. Their devotion did not alienate them from all professional care, and they did not consider it an impiety to conform to the rules of taste and compose a picture with which the eye should be charmed. Some indications show that, in the arrangement of their frescoes and their bas-reliefs, they had not always the deep intentions and mysterious designs attributed to them; that they simply let themselves be guided by reasons of order and symmetry; that they put certain subjects in certain places because they formed a pleasing spectacle, and that they placed opposite to each other scenes which, either on account of subject or date, ought to have been placed far apart; but the composition and arrangement of which seemed to have marked them out as pendants to each other.[2] Although antique art had placed itself so completely at the service of paganism, they studied its masterpieces and strove to imitate them. We have seen that in the early ages, when their faith was more fervent, they did not scruple to borrow from it images by which they represented their God. Of a truth, these loans never

---

[1] See, on these errors, Le Blant, *Étude sur les sarcophages d'Arles*, p. 8.
[2] Le Blant, *Sarcophages*, p. 13.

entirely ceased, and even in the works most directly inspired by the new religion, details are often found recalling the ancient legends and the art which so often reproduced them.[1] Thus, in becoming Christians, these artists did not abjure their understanding and love of the beautiful works of the painters and sculptors of Greece. They did not think themselves bound to condemn and proscribe them, since, on the contrary, they endeavoured to appropriate them to their religion. If it is true to say that the Renaissance had, above all, for its principle to clothe new ideas with the forms of ancient Art, the Renaissance began at the Catacombs.

IV.

THE CEMETERY OF CALIXTUS — SIGNOR ROSSI SUCCEEDS IN FINDING IT—THE INDICATIONS WHICH ENABLE HIM TO DISCOVER THE TOMBS OF THE MARTYRS—WORKS CARRIED OUT AFTER THE TIME OF CONSTANTINE IN THE CELEBRATED CRYPTS—*GRAFFITI* OF PILGRIMS—WHY THE CEMETERY TOOK THE NAME OF CALIXTUS—HISTORY OF THIS POPE, ACCORDING TO THE *PHILOSOPHUMENA*—WHY THE POPES OF THE THIRD CENTURY WERE BURIED IN THE CEMETERY OF CALIXTUS, AND HOW IT BECAME THE PROPERTY OF THE CHURCH—DISCOVERY OF THE PAPAL CRYPT.

WE have thus far been content to study the Catacombs in general: we have sought to ascertain their destination,

---

[1] Thus the monster which swallowed Jonah is represented just like the one which threatens Andromache; dead Lazarus is placed in a pagan *heroum;* and Noah's ark is an exact reproduction of the chest in which Danae is exposed on the waves, etc.

M

have described the aspect they present to the visitor, and have spoken of the inscriptions and paintings they contain. On all these subjects Signor Rossi has shed much light; but he has done more—or rather he has done something else. He continually repeats that his method is altogether analytic. He will not begin, like so many others, by general views, and, with him, generalities only result from the study of the details. It is these minute researches which he regards as the most important, and on which he especially prides himself. They must not, therefore, be forgotten when we undertake to make known his labours to the public. In order that their character and results may be quite understood, let us show him at work. By following him for a moment, and walking step for step behind him, we shall be able to understand the sureness of his method and the greatness of his discoveries.

Signor Rossi, being resolved to proceed systematically, decided to study the different Christian cemeteries in order of their importance. He should therefore have begun with the crypts of the Vatican. St Peter was buried there, and for two centuries his successors chose to rest near his tomb. But these crypts have been crushed, as it were, under the foundations of the immense basilica built above them, and nothing of them now remains. After the cemetery of the Vatican, which was inaccessible, hierarchical order designated that bearing the name of Calixtus, said to contain the sepulchres of the Popes of the third century. It is on this side that Signor Rossi directed his researches.

It was first of all necessary to find its site, which was no easy matter, for there was never a cemetery concerning whose position there had been so much discussion. That it must be along the Appian Way was well known; but some confounded it with the Catacomb of St Pretaxtatus, others with that of St Sebastian. In the latter, marble slabs were even placed, which still exist, solemnly informing visitors "that they were in the place where St Cecilia was interred, and where rest the fifty Popes," that is to say, in the cemetery of St Calixtus. But this bold assumption of possession did not intimidate Signor Rossi. These slabs were placed in position in the fifteenth century, that is to say, when the Catacombs had been almost forgotten, and in his researches Signor Rossi was resolved only to decide by means of documents, going back to the period when they were known and visited, and when the name of each was accurately known, together with the martyrs it contained. First among those documents must be put a species of writings whose whole importance had hitherto not been recognised. The ancients, like ourselves, had "guide-books," and, indeed, in a town like Rome, where all the world congregated, it would have been difficult to do without them. Those preserved to us belong to the last period of the Empire. Usually an enumeration of the wonders of Rome is found in them—the squares, the palaces, the theatres, the porticoes, etc. They also contain itineraries, such as are found in the guide-books of our time, in which the traveller is conducted from one end of Rome to the other, telling him all the buildings he will meet with on his way. The old editions of these itineraries are short and dry, but in the more recent ones the need of interesting the reader is felt, and he is told a crowd of extraordinary legends, in order to make him feel more pleasure in the curiosities he is shown. M. Jordan is even tempted to believe that they were sometimes adorned with illustrations, representing the more curious monuments;[1] so that nothing was wanting which conduces to the success of our own guide-books. Their use was still frequent, even in the Middle Ages, and we possess itineraries of the sixth and seventh centuries, which guided pilgrims to the tombs of the martyrs. They may be said to have rendered Signor Rossi the same service, and shown him the way among the most celebrated Catacombs. Two of these itineraries especially, discovered at Salzburg in 1777, enumerate with a sufficiency of detail the Catacombs of the Appian Way. It is, thanks to them, that Signor Rossi was enabled to find the site of the cemetery of Calixtus again.

The cemetery discovered, and the entrance to the underground galleries cleared, much still remained to be done. The itineraries informed Signor Rossi what tombs pilgrims of the seventh century went to visit there; but they still had to be found. This was not an easy work. How was he to know where he was, and find his way amid those hundreds of galleries and thousands of tombs? How could he be sure that he was taking the road leading to the famous crypts? Happily, here also precious indications were about to guide his researches.

These indications were furnished to Signor Rossi by the works carried out in the cemeteries at the peace-

---

[1] Jordan, *Topogr.*, I. 50.

ful period of the Church, and of which the remains are still easily distinguishable at the present time. Triumphant Christianity honoured the asylum of its evil days, but, as the Catacombs had suffered much during the persecutions, and everything could not be repaired, especial care was directed to the crypts where the chief martyrs reposed. They were strengthened and embellished; new and finer entrances and more convenient staircases to descend into them were constructed. Wells (*lucernaria*) were also dug to give them light. The poet Prudentius, who saw the Catacombs under Theodosius, has described for us, in beautiful verses, the state they were then in, and the ever-flowing stream of pilgrims who visited them.[1] He complacently depicts the holes made in the roof to light the more important crypts; he shows the darkness of the galleries interrupted from time to time by kind of islets of brightness, and those alternations of darkness and light which filled the soul with a religious awe. Near the tombs of the saints the walls are covered with marble, or lined with plates of silver "which shine like a mirror." It was thither people repaired from all sides when the feast of some famous martyr came round. They come from Rome, and the imperial city pours forth the flood of its citizens. "They come also from the neighbouring lands. The peasants hasten in a crowd from the villages of Etruria and the Sabine." Each one sets gaily forth with his children and his wife. They hie them forward as swiftly as they may. The fields are too small to hold this

---

[1] Prudentius, *Perist.*, XI. 155, etc.

joyous folk, and on the road, vast though it be, the huge crowd is seen to stop. "It is the same people who, in our days, willingly leaves its marshes or descends from its mountains in order to visit the miraculous Madonnas or the Bambino of Arc Cœli. Arrived at the tomb of the martyr, they all yield themselves up to that expressive and clamorous devotion of which the Italians have not lost the habit. From early morn they crowd to salute the saint. The throng who come to adore him pass and pass again till eventide. They kiss the shining silver plate which covers the tomb, they scatter perfumes there, and tears of tenderness flow from every eye."

The pilgrims spoken of by Prudentius have left in the Catacombs traces of their passage. They were in the habit of writing their names, with some prayer, along the staircases and at the entrance of the crypts. Time has not altogether effaced these *graffiti*, which are found especially near the most frequented tombs. Signor Rossi faithfully copied all that he was able to read, and his trouble was not wasted. How many curious particulars are revealed to us by those few words which rude peasants of the fifth or sixth century traced upon the walls! Among other strange revelations they acquaint us with one of those thousand links by which Christian devotion is attached to earlier beliefs. When we look from afar off, these delicate connections escape us, and it seems to us as if an abyss separated Christianity from the religions that preceded it; but science, which studies things closely and neglects no detail, without entirely bridging the distance, at least re-establishes the transitions. It was a pious custom of the Greeks and Romans, on visiting some famous temple, or even some monument that struck them with admiration, to recall their relations or their friends to mind, whether in order to recommend them to the god to whom the temple was consecrated, or to associate them in the pleasure produced in themselves by a fine sight. These acts of adoration or *proscynemes*, as they were called, in which the traveller associates with his personal impressions the names of those who are dear to him, occur frequently in Greece, and, above all, in Egypt. They are usually somewhat short, and little varied in their form. " Serapion, son of Aristommache, is come near unto the great Isis of Phile, and, moved by piety, remembered his parents." "I, Panollios of Heliopolis, have visited the tombs of the kings, and thought of all my kindred." Yet all are not so simple and so cold, and we sometimes find veritable feeling in them. A Roman lady, visiting the pyramids, remembers the brother she has lost, and writes these touching words: "I have seen the pyramids without thee, and the sight filled me with sadness. All I could do was to shed tears for thy fate; then, faithful to the memory of my grief, I was fain to write this plaint." Signor Rossi, therefore, was not quite right in saying that the pagan *proscyncmes* never contained anything more "than a cold and sterile formula," but Christianity certainly put more warmth and passion into them. What interests us above all in them is that they are natural and spontaneous. There is nothing official or conventional about them, as in the great inscriptions graven on marble. They are less pompous and less magnificent, but the impulse of the heart is much better felt. Sometimes the pilgrim simply writes his name, humbly asking some boon for himself, and uttering a few pious wishes for others.—*Eustathius, humilis peccator; tu qui legis, ora pro me, et habeas Dominum protectorem.* At others, he implores the saint for himself or for the persons whom he loves: "Holy Martyrs, remember Dionysius!" "Ask that Verecundus and his people may have a happy voyage." "Obtain repose for my father and for my brothers." But most often he is content to use this short formula, "Live!" or, "May he live in God!" At the entrance to the crypt of Lucina are found these words, several times repeated: "Sofronia, live in God!" (*Sofroniam vivas!*) Doubtless, after writing these words, the traveller went into the crypt, knelt down, and prayed at the foot of the martyr's tomb; and confidence entered his heart with prayer. This is proved by the following inscription, traced by the same hand on the exit side: "My dear Sofronia, thou shalt live for ever; yes, thou shalt live in the Lord!" (*Sofronia dulcis, semper vives Deo. Sofronia, vives!*)

Signor Rossi does not so carefully gather the mementoes, left by the period of Constantine and of Theodosius in the Catacombs, from mere curiosity. They have for him quite another importance, since they put him on the track of the historical crypts. As it was for them alone that, after the triumph of the Church, these broad staircases were built and these large light-holes dug, he is warned, when he comes across them, that some illustrious tomb is not far off. In order to find it, he has only to put himself in the footsteps of the pilgrims, of whom I have just spoken. Their *graffiti* guide him; he walks as it were with them, and by the increasing warmth of their prayers, he can guess that he is nearing his goal. Once in the crypt, a hundred details, which he carefully notes and compares with information furnished by the ancient historians, now informs him what martyr or confessor it is whose sepulchre they thus came to honour, and whose help they invoked. If it is a renowned saint, he rarely fails to discover, by diligent search, some remains of an inscription by St Damasus. This Pope was a great admirer, or rather a great devotee, of the Catacombs. He passed his life in repairing and embellishing them. He even composed little pieces of verse, to be placed above the tombs of saints, and recall their actions to the faithful. For their engraving on marble, Furius Filocalus, a renowned caligrapher, who styles himself the admirer and friend of Pope Damasus (*Damasi papae cultor atque amator*), imagined a special alphabet, whose letters have at their extremities certain ornaments which enable them to be easily recognised. As they were never used except for the verses of the Pope-poet, one is certain, on beholding one of these letters on a piece of broken marble, that one has in hand a fragment of an inscription by Damasus, and that one is consequently near the tomb of some great personage.

It is by such means that Signor Rossi succeeds in finding his way almost with certainty in this labyrinth, and how, in so few years, he has discovered again so many famous tombs. Yet there was a sepulchre which he still lacked, and it was just the one most necessary for him to discover. The opinions expressed by him respecting the position of the cemetery of St Calixtus had the disadvantage of newness. The many people who never pardon novelty are wrong. Under a government of priests, in a country where immobility was a physical need and a religious dogma, to change the least thing in received opinions was regarded as a crime. In order to obtain forgiveness for his innovations, to open the eyes of the incredulous, and to victoriously demonstrate that he was really in the cemetery of St Calixtus, it was requisite to find the sepulchre of the Popes of the third century. The question which Signor Rossi was trying to solve was full of obscurities. Why had this sepulchre of the Popes, which on the faith of ancient documents he was obstinately seeking in the cemetery of St Calixtus, been transported thither? How came it that the Bishops of Rome had chosen to rest elsewhere than beside St Peter, in the glorious crypt of the Vatican? No one had been able to find the reason. Nor was this the only subject of uncertainty and doubt presented by the study which Signor Rossi had undertaken. From the very beginning of his explorations, he perceived that the cemetery of Calixtus is much more ancient than the name under which it is known led one to believe. The character of the paintings in the chambers and galleries first excavated; the arrangement of the tombs, with the character of the inscriptions found in them, all recalled the second half of the second century. A still more decisive argument is, that the bricks used in their construction and which, according to Roman custom, bear the maker's mark, were all made under Marcus Aurelius. These works are therefore anterior to Zephyrimus and Calixtus, who lived under Severus. Certain indications seemed to prove to Signor Rossi that this first *hypogeum* really belonged to the second century, and that it was given to the Church by a member of the illustrious family of the Cæcilii. Why, then, did it not keep its first name, and how comes it that it bears that of Calixtus?

This is what we are beginning to know or foreshadow since the discovery and publication of a curious polemical work, written in the third century by an unknown theologian, and called the *Philosophumena*. This work, which, until our day, had remained hidden in the library of a Greek convent, caused, on its appearance, intense surprise and great scandal. It certainly very much upset accepted opinions. Above all, it described the life of this Calixtus, whom the faithful made Pope, and whom, later on, the Church made a saint, in a very unexpected manner. If the unknown author of *Philosophumena* is to be credited, this Pope and saint was merely a former slave who entered into banking with the money of his master, Carpophorous, and whom the too credulous Christians charged to keep the revenues of the Church. He succeeded badly in his operations, and dissipated the money which had been confided to him. In order to dispense himself from rendering his accounts, and by an exploit to reconquer his popularity, which his financial disasters had shaken, he bethought himself to go and make a noise in the synagogue of the Jews, and disturb their ceremonies. Exiled to Sardinia for this act of intolerance, and afterwards recalled to

Italy by the influence of Marcia, the mistress of Commodus, who protected the Christians, he became, it is not known how, the favourite and successor of Pope Zephyrinus. His character did not change with his fortune. He had been an unfaithful slave and a fraudulent banker: as Bishop of Rome, he was heretical, corrupt, simoniacal, "and by his example taught adultery and murder." This is certainly a not very edifying history for a Pope and a saint; and, happily, it is not at all credible. Signor Rossi has no difficulty in showing that the violence of this libel weakens its authority,[1] and that the accusations contained in it are entirely wanting in probability. The author himself is at pains to inform us that they are only an isolated protest, when he tells us that Calixtus has seduced everybody, and that he is alone in resisting him. But it is none the less certain that if, in writing for contemporaries, he perverted facts, he did not entirely imagine them. Signor Rossi thinks there must be truth at the bottom of the narrative, and that, for example, we must believe what it tells us of the origin of Calixtus and his first calling. So he was a former slave, and had long carried on banking on the Forum. Is it not significant that at this moment, scarcely two hundred years after the death of Christ, the Christian community of Rome, standing in need of a chief, should seek out an ex-banker? This means that it had already become rich, and was beginning to have a care for its temporal interests. It was no longer enough for him who directed it to be capable of govern-

---

[1] He has specially studied this question in his *Bullettino di archeologia cristiana* of 1866.

ing souls; he must also be capable of directing business. And it seems that, in choosing Calixtus, the Christians were not deceived. We gather from the involuntary avowals of the author of the *Philosophumena* that this Pope was a skilful organiser—a sort of liberal and enlightened man of State, who made useful regulations for the discipline of the Church. The people of Rome continued to remember his name long after the memory of his acts had been lost, and Signor Rossi rightly sees in this persistence a far-off remembrance of the great part which Calixtus had played.

We find in this violent pamphlet a very singular expression, which attracted Signor Rossi's attention first of all. It is said that, when Zephyrinus had been named Bishop of Rome, he sent for Calixtus from Antium, whither he had been exiled since his return from Sardinia, and gave him charge of *the cemetery*. The cemetery on the Appian Way which has kept his name, is doubtless the one in question. But how must we explain this strange manner of designating it? The Christians then possessed a great number. They had more ancient ones; for example, that of Domitilla, dating from the first century. They had more celebrated ones—as the crypt of the Vatican, where the first Popes were interred. Why, then, is the one on the Appian Way called *the cemetery*, as if it were alone? Signor Rossi, as we shall presently see, believes that the first *hypogea* possessed by the faithful were due to the liberality of some great lords converted to the new faith, and that, in the eyes of the law, they continued to be the property of the families who had conceded them; but he supposes that, later, the Christians profited by the protection which the emperors accorded to funeral associations, and succeeded in becoming legitimate and recognised proprietors of their burying-grounds, like the others. It is then probable that the cemetery of the Appian Way was the first, and perhaps for some time the only one, that enjoyed this privilege. Thence we understand how the ancient *hypogeum* of the Cæcilii, enlarged, embellished, and placed on a par with its new fortunes, should have become, above all others, *the cemetery*, and how they should have become accustomed to give it the name of Calixtus, who doubtless directed the works. And this is also why all the Bishops of Rome, from Zephyrinus downward, were buried there. They preferred the cemetery of Calixtus to all others, because it was the first of which the State had assured them the possession; they chose to be buried in the bosom of this earth which belonged to them, and in the domains of the Church.

And Signor Rossi believed himself certain of finding this sepulchre where they rested. Since ancient itineraries mentioned it, and pilgrims of the seventh century came to pray there, he could not fail to find it some day or another. He, in fact, succeeded in doing so in the month of March 1854, after five years of seeking, and by applying his ordinary methods. A considerable mass of ruins near the Appian Way had attracted his notice. And, sure enough, there was one of the large wells, or *lucernaria*, which had been dug after Constantine's time, in order to give light to the Catacombs. The workmen penetrated by this well into a chamber of medium extent (3.54 in length by 4.50 in breadth), but which must have been decorated with great magnificence. Successive restorations had covered the walls with delicate paintings, and then with slabs of marble. Unhappily, Signor Rossi had been forestalled in his researches. Devastators had got in, it is not known when, had finished the ruin begun by time, and, in order to obtain the marble, had destroyed a part of the inscriptions. But they had been unable to take all. The crypt being half-full of heaped-up materials, they could not reach the ground, and there was hope that among the rubbish which filled it some discovery might be made. So they set courageously to work to clear it out. As the walls reappeared, they were found covered with those *graffiti*, which never fail in important crypts. As always, they were the work of pilgrims addressing the martyr whose tomb they were visiting, and asking him for a happy passage for them and their families (*ut Verecundus cum suis bene naviget*). But who could this saint be to whom their prayers were addressed? It fortunately so happened that one of them had named him. In one of these inscriptions a name may be read, several times repeated: *Sancte Suste libera* . . . . *Sancte Suste, in mente habeas*, so that one of the greatest Popes of the third century was in question, beheaded in the Catacombs themselves, where he was celebrating the holy mysteries in spite of the Emperor's prohibition. They were therefore probably in the papal crypt where St Systus was buried with his colleagues after his martyrdom. But it was necessary to find more certain proofs. Signor Rossi has related with what anxiety he followed the work of his labourers, searching the rubbish as it was brought out of the crypt, without passing over the slightest fragments. At last, by putting pieces of broken marble together, he succeeded in recomposing the inscriptions placed on the tombs of four Popes. These epitaphs are remarkable for their simplicity. They contain neither praise nor regret, only these words are read: *Anteros bishop; Eutychianus, bishop.* On that of Fabian, another hand added, later on, the word "martyr."[1] No doubt was longer possible. All Signor Rossi's assertions found themselves confirmed by this brilliant discovery. It was indeed the papal crypt that had been found after fifteen centuries, and on the 11th May 1854, Pope Pius IX. came to visit the tomb of his distant predecessors.

V.

CHIEF RESULTS OF SIGNOR ROSSI'S DISCOVERIES — HIS NEW OPINIONS ON THE ORIGIN AND HISTORY OF THE CHRISTIAN CEMETERIES — THEY BEGIN BY BEING PRIVATE PROPERTY — AS SUCH THEY ARE UNDER THE PROTECTION OF THE LAW — HOW THEY EXTENDED — HOW THEY BECAME THE PROPERTY OF THE CHURCH — FIRST RELATIONS OF THE CHURCH WITH THE CIVIL AUTHORITY — CHARACTER OF THESE RELATIONS — THE PRIMITIVE CHURCH AND THE GREAT FAMILIES — HOW ADVANTAGE MAY BE DRAWN FROM THE ACTS OF THE MARTYRS.

We now know how Signor Rossi proceeds in the excavations which he undertakes. Seeing him working

---

[1] Signor Rossi thinks he may conclude from this that the title of martyr was only granted after a deliberation of the Church.

at St Calixtus, we understand his method. Instead of following him in the details of his other discoveries, I think it will be better, in finishing, to show the consequence which he drew from them. Of course, I do not pretend to enumerate all the obscure problems which he has solved; I limit myself to the most important. I will simply recall some of the new ideas with which he has enriched history, and the definitive conquests which Christian archæology owes to him.

First of all he has explained, better than has been done before him, the origin of the Christian cemeteries, and the phases through which they have passed. In this connection he has changed received opinions, and illumined with a new light that delicate question, the relations of the rising Church to the civil power.

In speaking of the Catacombs, underground places are usually imagined whose entrance is only known to a few initiated persons, and in which a proscribed religion carefully hides itself from its persecutors. This is an idea that must be got rid of—at least, as regards the two first centuries. It is now certain that, in the beginning, the Christians did not seek to conceal the existence of their cemeteries; that the authorities knew of them; and that until the persecution of Decius they never forbade access to them.

In the year 1864 the entrance to one of the oldest cemeteries of Rome was discovered—that of Domitilla. It was located along one of the most frequented ways— the *Via Ardeatina*. The door opened directly upon the road, and above the pediment is found the place of an inscription, which has disappeared, but which, as usual, told to whom the *hypogeum* belonged. Beyond the vestibule, a long gallery opens, whose roof is adorned with graceful paintings, representing a vine, with birds and genii. On the wall is seen the trace of more important frescoes, in one of which the subject of Daniel in the den of lions, afterwards to become so popular, is represented. All this first storey rose above the ground: it struck the eyes of all; it was impossible not to remark it. This is because the cemetery had at that time really nothing to hide. The person to whom it belonged, Domitilla or another, had a right to admit whom he chose to it. Have we not thousands of tombs whose owner tells us that he has built them for himself and his family, for his friends, for his freedmen and freedwomen, for those who form part of the same college? We have one even where he expressly mentions the people who belong to the same religion as himself, as destined to share his sepulture (*qui sint ad religionem pertinentes meam.*[1]) Signor Rossi, relying upon this usage, thinks that the Catacombs were originally the private tombs of rich Christians, and that instead of admitting their freedmen to them, they admitted their brethren. What lends sufficient probability to this opinion is the manner in which they are designated in the most ancient documents. They are usually called by a proper name, which is not that of martyrs or confessors buried in them. It is probably the name of the first proprietor of the tomb, of him who paid for the ground, and had the crypt constructed. Under these circumstances, it is easily conceivable that the construction of the first Catacombs should neither

---

[1] Rossi, *Bull. di arch. crist.*, 1865, No. 12.

have caused any surprise in the pagan world nor have been opposed by the powers. Pious women, who from the very first were the most fervent adepts of the new worship,—Domitilla, Lucina, Commodilla; rich and generous people like Calepodius, Pretextatus, or Thrason, had sumptuous tombs raised for themselves in advance; nothing was more natural; everybody did the same. They did not build them for themselves alone; this was still a very common usage. They chose to rest there with those who shared their beliefs: this was more unusual, but not without example. This tomb, into which so many people were received, did not, therefore, the less belong to Thrason or to Commodilla. It was still a private property, and, like others, was guaranteed by law. The respect of the Romans for tombs is well known. The spot where any person was buried, even a stranger or a slave, at once became sacred (*locus religiosus*). The law took it under its protection, and defended it from all outrage. The Christians profited by this protection, like everybody else; for there was no reason to deprive them of a common right. Even when, under Nero and Domitian, authority persecuted them, it is not seen that the persecution was extended to their cemeteries. The Roman law did not refuse sepulture to the criminals it had punished, and the tomb of a person who had been executed was as inviolable as the others.

Yet let us add that, even under these circumstances, the Christians were only sure of escaping lawsuits and chicanery on condition that the surface of the ground in which they dug their cemeteries belonged to them. The inalienable possession of the upper earth was the guarantee of the inviolability of the underground tombs. The law which declared sacred the spot where a man was buried, not only protected the tomb,—it also included its dependencies. These were regarded as inseparable from the tomb itself, and they profited by its privileges. Under the head *ground contiguous to the sepulchre* (*area cedens sepulchro*), they became, like it, inalienable. Now, these dependencies were often very considerable. The sumptuousness of their tombs was the first luxury of rich people. They loved, first of all, to surround the monument where they were to repose with a rather extensive space, where they had various edifices built, and which was sometimes bordered with large trees. Behind these trees, orchard, vines, and gardens extended, and often, behind the gardens, cultivated fields. They took great care to mark on their epitaphs the exact extent of the soil, which at times was no less than three *jugera*. They said that they reserved it for themselves alone; that they formally excepted it from their inheritance, and that they would not have it cut up or sold. If they had happened to have a vault constructed, they did not forget this circumstance, and we see a certain number of funeral inscriptions expressly mentioned among the things of which the deceased reserves to himself the indefinite possession—the monument and its *hypogeum* (*monumentum cum hypogæ*).

These usages afforded the Christians the opportunity to acquire the land needed for their burying-grounds, however extensive it might be, without causing surprise to any one. They also gave them the hope of possessing them always, without fear of its falling into profane hands. That they profited by them there can scarcely be a doubt. It may then be affirmed that, before constructing their crypts, they assured themselves the possession of the upper soil; that they made it, to use the consecrated term, "a ground contiguous to the sepulchre," and that, by some inscription which will perhaps be recovered, they placed the monument and its *hypogeum* under the guardianship of the law. Signor Rossi, in drawing up the plan of the different cemeteries, made an important observation. He remarks that, if reduced to their primitive elements, leaving out those works which are evidently posterior, only a few isolated groups remain, each of which forms a regular geometrical figure of slight extent. These respected limits, this anxious care taken to dig within a narrow space, instead of spreading at large, and this regularity of form which has been held to, are only satisfactorily explained by the assumption that in this underground work it has been sought not to overstep the bounds of a field possessed on the surface. Each of these groups is therefore the exact reproduction of that field, and they represent those little primitive *hypogea* given to the Church by rich protectors, or which it bought with its own funds. By transporting them, in thought, to the soil, and replacing the trees that were planted, and the funeral monuments that were built there, and by shutting them in with gravestones and walls, we may form some idea of those species of islets which the Christian cemeteries must have formed during the second century in the Roman Campagna, between the properties of the rich and the tombs of the different religions.

The primitive Catacombs had therefore but very slight extent, but ere long they necessarily grew. In the first galleries that were constructed, the niches for the dead were broad and far from each other, and there was much room lost. The number of the faithful continually augmenting, it soon became necessary to crowd the tombs and fill up the empty spaces. This means did not long suffice, and they were obliged to decide on the piercing of new galleries; but, in order to respect the law, care was taken not to exceed the limits of the field possessed, and there were sometimes as many as five stages of galleries superposed in the same crypt. The first was 7 or 8 mètres from the soil—the last attained a depth of 25 mètres. These enlargements of course yielded a great deal of room. According to Signor Rossi's calculations, ground with a length of but 125 Roman feet could furnish, with only three stories, nearly 700 mètres of galleries. It must have sufficed the community of Christians for some time. Yet, as the number of the faithful constantly increased, it at length became absolutely necessary to leave the ancient boundary, which would contain no more dead. These small *hypogea* were often near together; they pushed ramifications towards each other, and several of them, by joining, formed a cemetery. The cemeteries, therefore, are only the union of these originally isolated crypts, and if they have now such a large number of openings, it is because each crypt had its own, and kept it. Must we go further and believe with some of the learned that, later on, all these cemeteries united to compose a single underground Christendom? One would fain suppose so, since the imagination would be pleased by the idea that the faithful, who, during their lifetime so ardently aspired to form but a single fold, at least attained to this after their death; but it is impossible to believe it, since the nature of the soil opposed too many obstacles to such a union. The cemeteries are often separated one from the other by deep valleys and swamps, where the water collects after storms, and the galleries under these marshes would never have been practicable. The Christians well knew it, so they only constructed their cemeteries on the slopes of hills, and whatever desire we may suppose them to have to meet together after death, it is not possible to admit that they ever attempted to cross the valleys. After all, the Christian cemeteries, although separated one from the other, still present a unity of labours sufficiently grand to satisfy the most exigent of imaginations.

It is thus that, little by little, these primitive *hypogea* which the Church owed to the generosity of a few Christians, increased in size. In a century they took such vast proportions that it became difficult to continue to treat them in the same way, and for the law still to consider them as the property of families who had ceded them to the faithful. In fact, Signor Rossi thinks that they then changed their position, and in support of this he relies upon the following considerations. He bids us remark that Constantine, in the edict of Milan, orders to be returned to the Christians " The properties belonging, not to individuals, but to the entire community " (*ad jus corporis corum non hominum singulorum pertinentia*), and we know that the cemeteries formed part of those common properties which were restored to them. Before Constantine, therefore, the Church must have obtained from the Emperors the same privileges as the corporations recognised by the State, which had the right to possess property, and by this title have been the legitimate proprietor of its cemeteries. But at what point did it obtain this important right which the Emperors were so chary of granting? Doubtless, before the period of Decius and Valerian, when it was the object of such cruel persecutions. Now, in the reign of Severus, a notable change took place in Roman legislation, by which it seems natural that the Christians should have profited. The Empire, in the first and second centuries, was overspread with burying associations (*collegia funeraticia*). These were societies which, in return for the payment of a moderate sum monthly, undertook to provide their members with a proper burial and decent obsequies. The success of them is explained by the fear then felt that the soul would be wandering and miserable in the other life if the body did not rest in a fixed sepulchre, or if it had not been interred according to the rites. The Emperors, who in general mistrusted societies, and vouchsafed them but scant toleration, made an exception for these. Perhaps, as they were exclusively composed of poor people, they seemed to them less to be feared, and they hoped to increase their popularity by taking them under their protection. A *senatus consultum* authorised in advance all the burial societies that should be founded in the Empire, so that, in order to acquire legal existence, it was sufficient for them to get themselves inscribed under this heading in the registers of the magistrates. Once authorised, they had a right to possess a common exchequer, fed by the subscriptions of their members and the liberality of their protectors, and they could meet once a month for ordinary business, or as often as they chose, in order to celebrate the festivals of the association. It must be owned that this *senatus consultum* offered the Christians extraordinary facilities which must have greatly tempted them. It asked no sacrifice of their beliefs, it required no lie of them. This being so, Christians could well affirm that they also formed a "burial association," since they regarded it as their first duty to give an honourable sepulchre to their dead of every condition. In getting themselves recognised by the State, which could scarcely refuse them what it granted to every one, they not only became lawful proprietors of their cemeteries, but they acquired the right to meet together without being troubled, and to possess a common treasury. It was a great advantage; Tertullian's manner of expressing himself, the terms he uses in speaking of Christian associations,[1] and, still more, reason and good sense bid us believe that they would not voluntarily deprive themselves of it. If, indeed, the Christian community got itself accepted by the State as one of those *collegia funeraticia* which covered the Empire, the Bishop must naturally have been regarded as the responsible chief of the society, and doubtless, passed in the eyes of the magistrates for the president of the association. The deacon,

---

[1] Signor Rossi bids us remark that the expressions used by Tertullian in speaking of the quota collected each month in the assemblies of the Christians: *Modicam unusquisque stipem menstrua die opponit*, recall the terms of the *senatus consultum qui stipem menstruam conferre volent*, etc.

who was charged with the administration of the cemetery, played the part of the person who, under the name of *actor* or *syndicus*, managed common properties. It follows that the name of the bishop and that of the deacon were known to the authorities, who doubtless had frequent intercourse with them. It was necessary to inform them when the Bishop was dead, and to give them the name of him who had been appointed in his place. Signor Rossi even thinks he has discovered, by certain indications, that some of the lists of the Popes possessed by us come, not from the archives of the Church, but from those of the prefecture of Rome, where they were preserved with care, and whither the copyists went to seek them, in order to be sure of having an authentic document. Here, then, is the State for the first time in relationship with the Church, which had hitherto escaped it. They will henceforth take the habit of working together, and will unite so strongly that they will think themselves no longer able to separate and live without each other. We have arrived at the moment when those bonds are formed which will soon become so close, but it must be owned that, if the Church thought to gain greater security and more rest by these relations, it was mistaken. This protection which it asked from the State, and which it was so glad to have obtained, brought it little and cost it dear. Henceforth the Emperors know it better and have their hand more directly upon it; when they strike they will hit straight. Instead of deviating to the insignificant among the believers, they will unhesitatingly smite the chief of the community. They know his name and his abode, they seize him when they will, they exile or kill him according to their caprice, and, having got rid of him, they prevent another from being appointed. The position of the cemeteries is also changed. When they were a private property, and belonged, at least in appearance, to some great family, no one dared to touch them. But once become the common property of the Church, they followed its destiny. They were seized by the agents of the *fiscus*, pillaged by the soldiers of the Emperor, and the Christians were often obliged to destroy and fill them up themselves in order to save them from the ravages of the enemy.

Signor Rossi's manner of explaining the origin of the Catacombs and their legal position has the advantage of explaining facts which till now seemed very obscure. It was not understood how the Christians could accomplish such great works in their cemeteries, introduce their workmen to dig the galleries and take out the rubbish from them, without arousing the attention of the imperial police. Our surprise ceases from the moment we know that they did it quite openly, and with the assent of the authorities. This theory also allows of a better explanation, than was formerly the case, of the alternations which the Church passed through during the first two centuries. Its position was then a double one, and it was possible to be indulgent or severe towards it, according to the side from which it was regarded. As a new religion, it was to be interdicted. The law was precise, and forbade all foreign religions that had not been accepted by a decree of the Senate, but as "a funeral association" it was authorised. Hence a kind of hesitation of the authorities in their dealings with the Church, and the vicissitudes through which they caused it to pass. From time to time popular fury, always excited against the Christians, leads the magistrates of cities, the governors of provinces, and the Emperor himself to persecute the people who preach a new god. They had the legal right to do so, and, whatever apologists may say, the prosecutions are regular and lawful. But this effervescence of anger once calmed down, the rigours cease. It is affected no longer to regard the corporation of the "brethren," the adorers of the Word, as anything but one of those societies, half religious and half civil (*cultores Jovis, cultores Dianæ*, etc.) which have been instituted to give sepulture to their members, and they are allowed the same tolerance that is granted to others.

Signor Rossi bids us remark that this tolerance was rendered more easy by the care taken by the Church not to offend common usages when it found nothing to reprove in them, and to conform itself as much as possible to the customs of the ordinary associations. A pagan, who, in passing along the *Via Ardeatina* had been tempted to visit the cemetery of Domitilla, would have found nothing there to surprise him so much as we are inclined to believe. The charming arabesques adorning the roof of the entrance corridor, those vine branches so gracefully intertwined, those vintage scenes, and, elsewhere, those birds and those winged genii fluttering about in empty space, would have recalled to him what he had before his eyes every day in the apartments of rich people. The epitaphs, if he stopped to read them, might doubtless appear different enough from usual inscriptions; yet they contained hardly anything that might not be found elsewhere. Even the wishes, "Peace and refreshment," which seem to us the most original part of them, are borrowed from certain Oriental creeds that had been long since acclimatized in Rome. The Christian obsequies, too, at first sight, and to a hurried observer, must have greatly resembled others. Prudentius says that they bestrewed the tomb with leaves and flowers, and that they poured libations of perfumed wine upon the marble. Above all, the custom of celebrating funeral anniversaries by banquets had been preserved. Beside the entrance to the cemetery of Domitilla, the dining-room is still found, where the brothers met to celebrate the memory of their dead. Signor Rossi shows, by curious examples, how given they were to reproduce, at least to outward appearance, what passed in the *triclinia* of other associations; so that a pagan, assisting at these repasts, would have thought himself in one of those fine burial places possessed by great families and important associations along the Appian or the Latin Way. Other historians have been chiefly struck by the radical differences which separated Christianity from the different religions in the midst of which it established itself. Signor Rossi shows us the chance or intentional resemblances which it had to them. These resemblances made the transition from one religion to the other the more easy, which was doubtless not without profit to the rapid propagation of Christianity.

Another advantage of the explanations given by Signor Rossi is, that they enable us better to understand the relations of the first Christians to the authorities. People are fond of picturing Christianity to themselves as a kind of uncompromising sect which held civil society in horror, and would at no cost have anything to do with it; but there is a great deal of exaggeration in this opinion. The Church, on the contrary, during the first three centuries, made many efforts to live at peace with the ruling powers. Instead of putting itself in open revolt against the laws, it tried to benefit by those which were favourable to it, and even to gain admission to the class of the regular institutions of the Empire. These facts do not surprise us; we might have suspected them; but we had no such evident proofs as those which Signor Rossi gives us. Christianity is known to have been one of the rare Jewish sects of its age that were not at the same time a political insurrection and a religious reform. It declared from the very beginning that it could accommodate itself to all governments and live in all surroundings. Its founder preached submission to Cæsar in a country trembling with excitement and already almost in a state of rebellion. The Apostles, faithful to the doctrine of the Master, require obedience to all raised in authority. St Paul, in particular, seems to have taken much pains in order that the new religion might succeed in living with the old society. He does not choose that it should bring any dissension into the family or into the State; he forbids Christians who have infidel wives to separate from them, and he commands them to remain in the same position in which they were when they were called, and to maintain themselves in it before the Lord. His principle concerns the slave as well as the free man. They must all respect the social hierarchy, and render unto each his due. "Tribute to whom tribute is due, fear to whom fear!" They must, above all, be submissive to the prince, "who is the minister of God to favour the faithful in good." The Christians afterwards rigorously carried out these precepts of the Apostle. Even the persecutions themselves did not make rebels of them. In spite of the cruel manner in which they were treated, and which could not have disposed them to submission, we do not find them anywhere openly mixed up in the troubles of the Empire. Tertullian says that they prayed for the Emperor who persecuted them, and asked God to give him "a long life, a respected power, a happy family, valiant armies, a faithful Senate, an obedient people, and the repose of the universe." Signor Rossi renders these dispositions of the Christian community more evident, and makes us better understand the care which it took to avoid all conflicts and put itself straight with the powers, when he endeavours to prove that it profited by the privileges granted by the Empire to popular associations, and that it must have got itself authorized like other funeral associations, and kept up regular intercourse with the prefecture of Rome.[1]

He also introduced, in the history of the origin of Christianity, other opinions which were not quite ac-

---

[1] *Apropos* of the money which some churches consented to pay, in order to avoid the persecutions, Tertullian confirms that the Christians were on the registers of the police, and were there in very bad company: *Inter tabernarios et lanios et fures balneorum et alcones et lenones christiani quoque vectigales continentur.*—*De fuga in pers.*, XII. and XIII.

credited before his time, and which I will content myself
with rapidly pointing out. It has been often said
that Christianity at first only spread among the wretched
classes. It was poor Jews and "little Greeks," freedmen and slaves, weavers, shoemakers, and fullers, who
were its first converts. From the height of his rich
philosophy Celse laughed much at this mob of souls,
simple and ignorant, of minds narrow and unculti-
vated, for whom the Christian doctors mounted their
pulpits. It cannot, indeed, be denied that poor people
were for a long time the most numerous among the
faithful; but did the Church consist exclusively of
them, even in the first years? Signor Rossi thinks not.
He was struck to see that the most ancient Catacombs
are also the richest and best ornamented. He asks
himself whether it was possible for a corporation containing only "weavers and shoemakers" to produce
the vestibule of the cemetery of Domitilla, with the
elegant paintings that decorate the roof? And it at
once occurs to him that, among those slaves, those
freedmen, and those workmen, there must have been
more important and more opulent personages who bore
the expense of these buildings. This, however, is only
what happened in the poorer associations, which took
great care to choose protectors to help them with their
influence and their fortune. Is it not likely that something of the kind existed in the association of the
brethren? The excavations have seemed to confirm
these suppositions. On the tombs discovered by him,
Signor Rossi has occasionally read the most glorious
names of ancient Rome—the Cornelii, the Æmilii, the
Cæcillii, etc. He concludes that some members of

these great families very early knew and practised the new doctrine. Preached by St Paul "in the house of Cæsar," that is to say, among the Eastern slaves and freedmen of the prince, it had about the same time won over the noble Pomponia Græcina, wife of the Consul Plautius, the conqueror of Britain. She was accused, under Nero, of "foreign superstition," which could then only mean Judaism or Christianity, and as her descendants have been found in the cemetery of Calixtus, we may with much probability suppose that she was indeed a Christian. Some years later, the new faith made its way into the very family of the Emperors, if it be true, as there is every kind of reason to believe, that Domitilla and her husband Flavius Clemens, the nearest relations of Domitian and of Titus, were Christians, like Pomponia Græcina. Clemens and Domitilla could not have been alone, for an example in such high quarters is rarely unimitated by other personages. So it may be supposed that Christianity, even in its first years, made some important conquests among the aristocracy of birth and money which governed the Empire. Those great personages which it drew to it must, in the first place, have helped it by their position, and perhaps they more than once stopped the blows that were being prepared for it, like that Marcia, mistress of Commodus, "who feared the Lord," and protected the bishops. Above all, they must, by their liberalities, have enriched the common chest, which, from the period of the Antonines, was so important, and soon allowed the Church of Rome to extend its alms over nearly the whole world. The Catacombs have already revealed to us the names of some of those great lords who early became Christians, and when there was peril in so doing, and they will doubtless acquaint us with many others. They were perhaps a somewhat feeble element in that budding society, but one that must be taken into account. If we neglect it, it is less easy to understand how Christianity sustained the attacks of its enemies, and succeeded in vanquishing them.

Another question, perhaps more important yet, which is far from being settled, but which the study of the Catacombs has made a little clearer, is that of the amount of confidence deserved by the "Lives of the Saints" and the "Acts of the Martyrs." These documents are much discredited, not only among sceptics, but among pious people, like Tillemont, when they do not think that devotion renders abstention from criticism a duty. In the form that they have reached us, they deserve very little credence. In the centuries following the peace of the Church ridiculous legends got mixed up with them. When read on the *feasts* of the saints for the edification of the faithful, anything that could strike the imagination or touch the heart was added without scruple. Rhetoric, above all—the bad rhetoric of the seventh and eighth centuries—has quite spoilt them. Yet it must be owned that whatever distrust they may awaken, they must not, after the excavation of the Catacombs, be any longer rejected without examination. All in these narratives is not imaginary, since the graves of those whose history they narrate have been found in the galleries of the cemeteries. So, in the fourth century, their tombs were really believed to be extant; their names were read on their epitaphs, and people came to pray before their remains. The account of the facts may be very legendary, but it is difficult to doubt the reality of the names of the personages. Even in these very narratives, in the midst of many ridiculous errors, probable or certain details are remarked. Some are confirmed by the ancient inscriptions or paintings in the Catacombs; others imply a perfect knowledge of places which people of the eighth or ninth centuries certainly visited no longer. Signor Rossi very legitimately concludes from this, that the new, amplified and corrupted edition supposes the existence of an edition more ancient, more sober, and more truthful. So he is of opinion that, instead of rejecting the whole narrative on account of a few absurdities contained in it, we should strip it of all these lamentable accretions, and try to find the original truth under the sophisticated copy. It is delicate work, into which there always enters a little divination and hypothesis, yet one not impossible to a practised criticism, and which is accomplished every day in the restitution of classic texts. Signor Rossi has done it very cleverly for the "Acts of St Cecilia," and M. Le Blant is attempting it for many others. If, as is scarcely doubtful, the undertaking succeeds, it will greatly increase the number of documents at our disposal, and make us better acquainted with the heroic struggle sustained by the Church against its persecutors. We shall perhaps thereby gain a few additional martyrs, but this does not appear to me so very great an evil. I must own that I have never been able to understand the animosity with which historians of the eighteenth century systematically denied the persecutions, or sought to attenuate their effects. Voltaire, in treating the martyrs as enemies, did not perceive that he struck at allies. The men whom he pursued with his implacable railleries defended tolerance like himself. They proclaimed, like him, that no human power can touch the independence of the soul.

Come, tormentor," Prudentius makes a Christian girl say, "burn and tear! Divide the members formed of dust!" 'Tis easy for thee to destroy this frail assemblage. As for my soul, in spite of all thy tortures, thou shalt not reach it."[1] And, indeed, they did not reach it. Executions were useless, and Christianity gave the world the most moral of all spectacles,—that of the powerlessness of force.

The Church is indeed right to honour the memories of those who died for her, and to glory in their courage; but they are not the only heroes of a particular opinion. All those who think, like them, that belief must be free, and that a religion has not the right to impose itself by force, may shelter themselves under their name. It is therefore by no means to our interest to limit the number of the martyrs or to contest their merits; nor is it to our purpose to throw a shadow over that heroic epoch which gave so great an example to the world, and those who, like Signor Rossi, seek to make us better acquainted with it, whatever their personal convictions may be, have a right to the sympathies of

---

[1] Prudentius, *Perist.*, III. 90.

all. We should earnestly hope that the excavations directed by him will always be as productive, and that he will have time to finish this work, so valiantly begun. Were he to give us a few more saints and confessors than were acknowledged by Tillemont, we should have no reason to complain. By multiplying the victims he renders the executions more hateful; makes us the more detest that insolent intervention of force, which pretends to rule and regulate faith, and causes us to be more attached to those precious treasures conquered at the price of so many sufferings—Tolerance and Liberty.

CHAPTER IV.

HADRIAN'S VILLA.

No one who sojourns for any length of time in Rome fails to go and see Tivoli: the Cascatelle and the Temple of the Sibyl are almost as well known as the Colosseum or the Pantheon. Yet very few consent to turn aside for a moment from the accustomed track, to visit on their way what remains of the Tiburtine villa built by the Emperor Hadrian. Nevertheless, it is an excursion worth making, and one which can teach much to the lovers of antiquity. The monuments of Rome enable us to behold the Cæsars in the exercise of their sovereign functions, and preserve the memory of their official life. Hadrian's villa shows them to us during those moments of distraction and repose, which must needs be taken from time to time when one has the world to govern. It may also give us some precious hints as to their notions respecting the pleasures of the country, and inform us how that society understood and enjoyed nature—a subject well deserving of a moment's study.

On our way from Rome to Tivoli, we first pass, in all

its length, that desolate *Campagna* by which the Eternal City is on every side surrounded. After five or six leagues of veritable desert, where nothing is met with but a few wretched *osterie*, and herds of oxen or horses grazing on the scant grass, the ground begins to rise. Some clumps of trees announce the approach of the Anio, which we pass by the *ponte Lucano*. At this spot an ancient ruin of great interest rises, the tomb of the Plautian family. Here lies buried the Consul Ti. Plautius Silvanus, one of those brave officers and intelligent administrators who maintained the honour of the Empire under the worst princes, and were the salvation of Rome. The inscription placed on the mausoleum contains the account of his services, and the enumeration of the honours obtained by him. Under Tiberius, he commanded a legion of the army of Germany: he accompanied Claudius in the expedition to Britain, under Nero: he governed Mœsia, one of the provinces most threatened by the barbarians. The inscription relates how he stopped an insurrection of the Sarmatians, obliging the hostile kings to cross the Danube and come to his camp and bow down before the Roman eagles. These services met with but scant requital until the day when Vespasian, himself an old soldier, set to work to make good to his comrades the injustices of preceding reigns. He recalled Silvanus from his province, had him granted the triumphal decorations, and named him prefect of Rome.

From the tomb of Silvanus the road divides. To the left it enters those delightful olive woods that lead to Tivoli; to the right it crosses the plain, and brings one in twenty minutes to Hadrian's villa.

To-day that villa is little more than a heap of ruins. Over an extent of several kilomètres nothing is met with but immense substructions, shafts of columns, great scattered blocks, with here and there some fragments of wall still upright. These ruins are so considerable that they were for a long time taken for the remains of a town. It was imagined that Tibur, before climbing the hill, had been built in the plain, and that here one had the last vestiges of the ancient city before one's eyes. So in the country round about they had received the name of *Tivoli vecchio*. It was easy to show this to be a mistake: the testimony of ancient authors and the inscriptions on tiles proved that it was Hadrian's villa. This country-house, considered as a wonder by contemporaries, and which was the favourite creation of an Emperor friendly to the arts, does not appear to have been much dwelt in by his successors. History, at least, says nothing about it, and scarcely anything has been found in these ruins attributable to another epoch. It has therefore had the rare good fortune not to have been too much modified, and to have passed through the centuries bearing the special mark of the prince who built it, and of the epoch in which it was raised. The riches of all kinds found in the rubbish led to the supposition that it had not been pillaged so long as the Empire lasted: yet it must doubtless have greatly suffered when Totila ravaged the environs of Tibur, took the town by assault, and massacred its inhabitants. From that moment, its fate was sealed. The great halls fell in the plough passed over the avenues, and the gardens became cornfields. Nevertheless, in the fifteenth century, important ruins still remained. The illustrious Pope Pius II., who visited it, speaks admiringly of the vaults of the temples, of the columns of the peristyles, of the porticoes, and of the *piscinæ*, which could still be distinguished. "Age deforms all things sadly," he added. "The ivy climbs to-day along these walls, covered of yore with paintings and with golden stuffs; brambles and thorns grow where sat the tribunes clad in purple, and snakes inhabit the chambers of the *principes*. Such is the fortune of mortal things!" Even these ruins themselves were fated to disappear. To Hadrian's villa, as to most ancient monuments, the Renaissance was more fatal than barbarism. During the Middle Ages it had been allowed to decay; from the sixteenth century it was systematically destroyed. According to custom, excavations were made to search for the statues, the mosaics, and the paintings which it might contain, and in these hunts the walls that still remained standing finally fell in. Unfortunately for it, Hadrian's villa turned out to be much richer in such things than all the other ruins that had been excavated. It became for three centuries a kind of inexhaustible mine which furnished masterpieces to all the museums of the world. Thence, for example, came the Faun in *rosso antico*, the Centaurs in grey marble, and the Harpocrates of the Capitol; the Muses and the Flora of the Vatican, the bas-relief of Antinoüs of the villa Albani, and the admirable mosaic of the doves, so often reproduced by modern art. Of course, an edifice whence so many marvels were drawn was more conscientiously devastated than all the others. The pillage lasted down to our days, and, only a few years since, the Braschi family, who possessed a part of the ground, made over the right to exploit these ruins to a company, and how the company, who wished to recoup itself as soon as possible, went to work, may be imagined. Happily, the Government has put an end to this scandal by purchasing the villa Braschi.

In the state into which all these devastations have made it, Hadrian's villa is, for most visitors, a riddle, and if archæologists and architects did not come to our aid, it would be extremely difficult for us to find our way among these heaped-up ruins. Archæology has long been working to find out the destination of these blocks of stone and these masses of bricks, and to give us a plan more or less exact of the imperial dwelling. The first who busied himself with some success to this end was a Neapolitan architect of the fifteenth century, the famous Pirro Ligorio, the same who made himself so bad a reputation among the epigraphists by inventing entire volumes of false inscriptions. This great forger was certainly a very clever man. In his works on Hadrian's villa, he gave proof of great sagacity, and most of his conjectures have been adopted by the scholars who followed him. Piranesi and Canina have done little else than develop his views and exaggerate his errors. Nibby, who came after, contented himself with choosing the most plausible opinions set forth by others before him, and supporting them by his knowledge of texts and great experience of antiquities. The interesting book published by him in the year 1827, under the title *Descrizione della villa Adriana*,[1] might

---

[1] *Essai de restauration de la villa d'Hadrian.*

have passed for the last word that science had to say on the subject, when new studies were undertaken by one of the most distinguished architects of the French school at Rome, M. Daumet. In order to be more sure of the accuracy of his work, M. Daumet began by circumscribing it. He only busied himself with a portion of the villa—that which was known as the imperial palace. It offers many difficulties for solution, but also preserves the most curious remains. M. Daumet studied its least fragments with the greatest care; made excavations, when allowed to do so, endeavoured to ascertain the meaning of the smallest layers of stone, and put back in their places all the pieces of marble ornament or mosaic that he could find. The result of all these labours was a restoration essay on Hadrian's villa, considered one of the best and most complete works of the school at Rome. The excavations made since 1870, which have been very incomplete and intermittent, have sometimes confirmed M. Daumet's opinions and sometimes contradicted them. The work is far from being complete, and claims the expenditure of much additional time and effort, but, while awaiting its completion and the entire clearing of these ruins, it is, I think, useful to give an idea of what the works, carried out for three centuries past by distinguished archæologists and architects, have taught us most worthy of belief concerning this great curiosity of the past.

I.

THE EMPEROR HADRIAN — THE DIFFERENT JUDGMENTS PASSED ON HIM—THE PRINCE AND THE MAN—THE REASONS WHY HE WAS NOT LOVED — HIS LIKING FOR THE GREEKS—TRAVELLING IN ANCIENT TIMES —HADRIAN'S JOURNEYINGS.

HADRIAN'S villa has the peculiar characteristic of being the work and the personal conception of a man who was one of the most curious figures of his time : it was born of certain circumstances of his life, and bears everywhere the impress of his mind. We can only hope to understand it by first becoming acquainted with him who caused it to be built. We must therefore study the artist before the work, and strive to know both what he was and how the thought occurred to him to construct this country house, which filled all his contemporaries with wonder.

The Emperor Hadrian was descended from an Italian family that had been for a long time settled in Spain. His birth did not seem to predestine him for the Empire. He was half-cousin to Trajan who, after much hesitation, ended by adopting him on his death-bed. It was singularly fortunate for the Roman Empire that Nerva and the three princes who came after him were without male heirs, and were obliged to provide themselves successors by adoption. This absence of direct heredity is usually regarded in monarchies as the greatest of misfortunes, and it is now a principle adopted by all that, in order to insure the security of states, it is good for the son to succeed his father. The Romans had very different ideas, and even under the Empire preserved the remains of republican prejudices which rendered them little favourable to hereditary royalty. Their experience of it under the Cæsars and the Flavii had not reconciled them to it. After the fall of Domitian, they declared that they would not be the heritage of a family. "It appeared to them better for a prince to elect his successor than for him to be received at the hands of nature." "To be born of royal blood," said Tacitus, "is a chance before which all examination stops. On the contrary, he who adopts is judge of what he does; if he desires to choose the most worthy, he has only to listen to the public voice."[1] What is certain is that adoption gave the world a succession of four great princes, and that Rome was quite happy until the day when Marcus Aurelius was so ill-fated as to have a son and leave him the Empire.

I have just unhesitatingly put Hadrian among the great emperors, beside Trajan and Marcus Aurelius, but this is not, however, the opinion of all historians. His reputation is not one of those concerning which perfect agreement has been arrived at, and there are great differences in the manner of judging him. These difficulties go back very far—to the very epoch in which Hadrian lived, and his contemporaries probably agreed no better concerning him than we do. Dion and Spartianus, the chroniclers who have related his life, speak of him in a very odd manner. They at the same time say a great deal of good and of bad of him, so that the means of attacking or defending him may both be

---

[1] Tacitus, *Hist.*, I. 16.

drawn from their works. This is because he was really a very complicated being—*varius, multiplex, multiformis,* says his historian, gentle or severe, according to the occasion, economical and prodigal by turns, jocose or grave, a good-natured friend or a cruel railler. His life contained contrasts which could not be explained. Although an excellent general, he detested war, and always avoided it. He passed his time in exercising his legions, that were never to be led before the enemy. This scholar, this artist, this squeamish exquisite, did not hesitate when needful to enter into the most trivial details of common business, and the effeminate fop, who made little verses on tooth-powder, was capable of the most energetic resolutions. He had sumptuous palaces built for him, in which all the elegancies of luxury and all the refinements of comfort had been brought together, yet he willingly lived in a tent, satisfied with bacon and cheese, like the common soldiers, drinking only vinegar mixed with water, and marching bareheaded at the head of his troops, amid the snows of Britain and under the Egyptian sun. It is easy to understand that these contrasts should have troubled chroniclers not endowed with a very perspicacious mind, and that, in the presence of a prince in whom contrary qualities seemed to unite, they should have swayed undecidedly between opposite opinions, unable to take upon themselves to give us a precise idea of him.

What stands out most clearly in their accounts is that there were in Hadrian two personages who did not always agree very well together—the man and the emperor. The emperor only deserves praise, and may be placed among the greatest and the best: the man, on the contrary, was often unpleasant and petty. Contemporaries, who were placed too near, and did not always know how properly to distinguish, sometimes by their unjust judgments made the prince pay for the caprices and weaknesses of the man.

They were assuredly wrong, and all their gossip must not prevent us from believing Hadrian to have been a great prince. If any doubt still remained on this point, I should appeal to the brilliant picture of his reign recently drawn by M. Duruy.[1] The services of every kind rendered by Hadrian to the Empire are eminent and incontestable. He first gave his states external security, and, in order to maintain the discipline of the armies, he made regulations so wide that no need was ever found to change anything in them, and they lasted as long as the Roman domination. He strengthened the frontiers by garrisoning them with troops and by furnishing them with formidable retrenchments, thus shutting the door against the barbarians, who were becoming more formidable every day. Within this belt of ramparts, of fortresses, of deep dykes and of entrenched camps, skilfully disposed along its immense

---

[1] In the third volume of his *Histoire des Romains*. I am happy to refer the reader to this work, in which M. Duruy, returning to the labours of his youth, after a long interval and important services rendered to the country, sketches, in a manner at once learned and lively, the history of the Empire. Even when he does not succeed in quite converting the reader to his ideas, he still knows how to interest and instruct him. M. Renan, in the sixth volume of his *Histoire des origines du christianisme*, also speaks of Hadrian. He does not dissimilate his faults, but he at the same time gives prominence to his great qualities, and has traced of the prince one of those portraits which are not forgotten.

frontiers, the Empire could breathe in peace. Inside, tranquillity was maintained with a firm hand, abuses reformed, legislation softened, and a great impetus everywhere given to public works. Under this vigorous impulsion, and, thanks to the peace enjoyed by the world, the towns would adorn themselves with those splendid monuments which still excite our admiration. Thus much is undeniable. Hadrian was certainly one of the most able administrators that had governed the world since Augustus, and he perhaps contributed, more than anybody, to that incredible development of the public prosperity which made the century of the Antonines one of the happiest periods of humanity. "When the glory of princes is measured by the happiness they have given their peoples," says M. Duruy, "Hadrian will be first of the Roman Emperors."

How comes it that, having served the Empire so well, he was so unfavourably judged? These severities of opinion are usually explained by recalling the persistent bad humour with which the great families and the Senate regarded the imperial rule; but this is truly a somewhat too convenient means of justifying all the Cæsars without distinction, and, even if such reasons might still serve for the period of Nero and Tiberius, it seems to me scarcely possible to continue to use them when we get to the Antonines. The Empire had then long since been excepted by all. Time had wakened old republican rumours, and, at any rate, it can scarcely be understood how, after Trajan had been spared them, they should have been resuscitated against Hadrian. If Hadrian, with all his great qualities, could not make himself better loved, we must think it his own fault, and that there was in his person and in his character something which estranged hearts from him. So Fronto, who was a rather bad writer, but a very good sort of a man, and the most submissive of subjects, later on gave Marcus Aurelius to understand, with an infinity of delicate precaution. "In order to love anybody," he said, "we must be able to approach him with confidence, and feel at our ease with him. This is what didn't happen to me with Hadrian. Confidence failed me, and the very respect with which I was inspired repelled affection."[1] We may see all that is hidden beneath those polite words. Neither does Trajan, although his kinsman, seem to have felt any great attraction towards him. Yet we know that Hadrian, who expected everything at his hands, neglected nothing in order to please him. He sought to flatter his tastes in every way, even the least honourable, and he himself relates that, knowing him to be a hardened drinker, he set himself to drink, in order by this means to get into his good graces. Moreover, he had other qualities to which Trajan attached the highest importance. A devoted soldier, an exact lieutenant, a skilful organizer, a scrupulous administrator, he accomplished carefully and with success all the missions with which he was charged. Yet his advancement was not very rapid. An inscription, found in the theatre of Athens, shows that he went through the whole hierarchy of public dignities, step by step, without being spared a single grade. In spite of his acknowledged merits and the services rendered by

---

[1] Fronto, *Ad. M. Cæs.*, II. 1 (p. 25, Naber).

him, Trajan waited until his last day before adopting him. It is even pretended that death forestalled him before he had come to a decision; that the adoption was only a comedy scene, imagined in order to deceive the world, and that a man concealed behind the hangings murmured a few words in a dying voice, in place of the deceased Emperor. What might give some probability to such a tale is Trajan's apparently slight alacrity to accept him as his heir. Not only did he not associate him in the Empire during his lifetime, as Nerva had done for himself, but he would not confer upon him any of those exceptional honours which would have designated him in advance as his successor. May it not hence be concluded, that, while appreciating in him the administrator and the soldier, he felt for the man a sort of repugnance, which he had difficulty in overcoming?

Once Emperor, Hadrian had many friends; it is not difficult to have them when one is master of the world. He was very liberal towards them. "Never," says Spartianus, "did he refuse what they asked, and he often even forestalled their desires," but, at the same time, he irritated them by his railleries and wounded them by his suspicions. Unequal and fantastic as an artist, easily set against those who were attached to him, he listened to what was said against them, and often had them watched. He had his secret police, who penetrated into families and reported to him what it had heard said. No friendship whatsoever is proof against such mistrust. Spartianus remarks that those whom he had best loved and most loaded with honours all ended by becoming hateful to him. Several were sent away from Rome, others lost their fortunes, and there were some among them whose lives were taken. I do not believe that Hadrian was by nature cruel, and he even gave some fine examples of clemency. But it has been said that the sovereign power, devoid of any precise character and without fixed limit, would trouble the best heads. Few princes have quite escaped the intoxication of authority—that giddiness produced at once by pride and fear, which influenced bad instincts and perverted their souls. Honest Mark Aurelius one day said to himself in a tone of terror: "Become not too much Cæsar." We are forced to believe that Hadrian occasionally became so, in spite of himself. At the beginning of his reign, ere yet he felt himself firmly settled, he shed, or caused to be shed, the blood of some great personages accused of treason. He shed it again at the end of his life, and this time his brother-in-law, an old man of ninety years, was among the victims, and his nephew, who was not yet twenty. I am willing to believe that they were both guilty, and that the Emperor believed his rigour necessary, yet public opinion revolted at it. It was remembered that Trajan, to whom the Senate solemnly decreed the surname of excellent prince (*optimus princeps*), never had recourse to such lamentable necessities, and it was found that Hadrian resigned himself to them too readily. These executions, ordered by a dying prince, like a last rancour which he desired to sate, made good people indignant "He died," says Spartianus, "hated by all."

I know that the enemies of sentimental politics will maintain that it was wrong to hate him. It will be said that these family strifes scarcely interest the world, and that too much importance must not be attached to them. What matters it to obscure citizens, who form the great majority of a country, that a prince be unpleasant and make those about him suffer? If he governs his state well, if he preserves it from outside enemies, if he gives it peace within, ought we not to shut our eyes to his caprices, and allow him to deliver himself as he chooses of the friends who bore, and the relations who inconvenience him? What evil comes of it to his people? Certainly, if subjects were reasonable, they would judge their sovereign by the good he does to all, and not by the severity which only affects a few persons, and he would seem to them most worthy to be loved who made the happiness of the largest number. But love does not always reason, and affection includes other elements besides interest. So it is not uncommon to see sovereigns, under whose rule it is advantageous to live, who do not succeed in making themselves beloved. Hadrian was of the number. Even at this distance of time, we cannot quite overcome the sentiments with which he inspired people of his time, and we are obliged to make a kind of effort over ourselves, in order to esteem him as much as he deserves. However convincingly M. Duruy may prove to us that he rendered more service to the world than Trajan or Marcus Aurelius, we shall find it difficult to blame his contemporaries who loved Marcus Aurelius and Trajan better than him.

To these general reasons which the Romans might have for not loving him, may be added others more peculiar to themselves. Perhaps a little resentment may have entered into their severity against a prince who took pleasure in braving their prejudices, and who openly sacrificed them to their eternal enemies. The influence of Greece was then stronger than ever in Rome. It at the same time seized society by two extreme points. Upon the rich, upon the great lords, and upon people of the world it forced itself by education and by the sovereign charm of the Arts and of Letters. In those sumptuous palaces of the Esquiline, in those magnificent villas of Tusculum or Tibur, where reproductions of the masterpieces of Praxiteles or of Lysippus courted the eye, and where Menander and Anacreon were read with such pleasure, people had become more than half Greek. In the popular quarters they had grown so entirely. There a continuous immigration brought from all parts of the Orient people who found it difficult to live at home. It was a stream which, for many centuries, had never stopped. What would old Cato have said, could he have seen Greece and the East thus established on the Aventine, and that race he so despised nearly mistress of Rome? It was shame and danger that faced the old Romans, and they naturally found that it was an emperor's duty to combat them.[1]

Hadrian, on the contrary, put himself on the side of the Greeks. He had devoured their great writers from his earliest years, and so delighted in their language that it became difficult for him to speak any other. One day,

---

[1] The violent expression of these sentiments will be found in Juvenal's third satire.

when, in his capacity of quæstor, he had to read a message from Trajan, he spoke Latin so badly that the Senate laughed at him. To admire Greek art did not suffice him; he chose to be an artist himself, and in every branch of the art. He became at the same time a musician, a sculptor, a painter, and an architect; he prided himself on singing well; he danced with grace; and he knew geometry, astrology, and enough of medicine to invent a *collyrium* and an antidote. The Greeks had not praises sufficiently hyperbolical for a prince who excelled in so many different things; but the Romans, on the contrary, were inclined to laugh at him. The more sensible owned that it is certainly not a crime to know how to make statues and to paint; but they added, neither is it of much use, when one has the world to govern. It seemed to them that this great business admitted of no other in addition, and that it claimed the whole of a prince's activity. They remembered, too, that those Emperors who had too much loved the Greeks, and who had made it their glory to imitate their customs and win their praise—Nero and Domitian, for example—had been abominable tyrants, and these recollections were not calculated to make them favourable to Hadrian's hobbies.

What irritated them still more was to see the important part taken by Greece in the political affairs of Rome. For a long time she had been content to rule in intellectual matters, and to furnish Rome with grammarians and artists; but from Hadrian's time she openly invades what had hitherto seemed forbidden to her and reserved for the victorious race. She slips into the armies, takes a place in the Senate, and governs the provinces. We see among the generals of that period the names Arrienos and Xenophon. Of course the Greeks were very much flattered at this. Their gratitude knew no bounds, and, in accordance with their wont, they expressed it in a base and servile manner. In their most important cities magnificent temples arose in honour "of the new Jupiter, of the Olympian god," and his worthless favourite, the beautiful Antinoüs, also a Greek, received after his death the most extravagant honours. But it is not unnatural that the old Romans who remained should have been indignant. It will perhaps be said that they were wrong, and that there was nothing in Hadrian's conduct to excite surprise, or anything contrary to the institutions and the principle of the Empire. The Empire having called upon the provinces to share the sovereign authority, the turn of Greece and of the East must one day come, and it was not very surprising to see Greek generals and proconsuls under Spanish Emperors. But a distinction must be made. While the provincials of the West, admitted by Rome into its armies and destined for public dignities, adopted the language and the customs of their new country, assumed its spirit and its ancient maxims, and became frankly Roman, the Greeks remained Greek. Nothing could ever modify this supple and tough race, which passed unchanged through the Roman domination, and survived it. It kept its pride even in its servility, and, while flattering the barbarians, frankly despised them. So it had no difficulty in keeping itself from all imitation of their customs, or from fusion with them. I do not believe that any Greek ever became quite Roman, but, on the other hand, many Romans became entirely Greek. Even in Hadrian's time we see Favorinus the Gaul, who was born at Arles, and the Italian Elienus of Præneste, abandoning their native language for that of Greece. That this invasion of a foreign spirit should have wounded serious Romans cannot excite surprise. They were quite right in thinking that Rome had everything to lose by it. The different nations who entered into the Roman unit brought their national qualities and rejuvenated the Empire, whereas the Greeks only communicated their faults to it. By favouring the invasion of this new spirit, therefore, Hadrian was at least guilty of imprudence; he unconsciously worked at hastening the hour of the lower Empire.

Such, with his singular admixture of great qualities and defects, was this Emperor, half Roman and half Greek, the originator, and perhaps even the architect of the villa at Tibur. It remains for us to ascertain the occasion of his building it. Historians tell us that, at least, the greater part was constructed in consequence of his travels, and in order to preserve their memory. It is known that Hadrian lived very little in his capital, and passed nearly all his reign in travelling over his vast Empire. Nothing so much struck the world as this active life and these endless journeyings. The populations who saw him so often go by, retained the memory of an indefatigable traveller who was unceasingly passing from one end of the universe to the other. "There never was a prince," says his biographer, "who so rapidly visited so many different countries."

Not that travelling was then so uncommon as is usually supposed. People no more liked stopping in

one place in ancient times than they do in our
own days. Seneca was so struck by this craving for
movement and change of place by which men are
tormented, that he tried to give a philosophical explanation of it. He attributes its origin to that divine
part which is in us, and which comes to us from the
stars and the sky. "It is the nature of celestial
things," he says, "to be always in motion."[1] Since the
Empire had given peace to the world, travelling, being
safer, had also become more frequent. Those narrow
roads, solidly paved with large slabs, which led from
Rome to the ends of the world, were constantly
traversed by the chariots of knights and by pedestrians.
People of all fortunes were seen to pass along, from
him who, like Horace, only mounted a poor mule, short
of tail and heavy of gait, to those great lords stretched
in their comfortable litters, where one could read,
write, sleep, and play at dice, preceded by Lybian
couriers and followed by a whole train of slaves and
clients. All these people found more facilities for
making the journey than we are inclined to think.
The Imperial post had just been established, and
provided all those furnished with an authorization
from the Emperor with horses and carriages, which
made about 8 kilomètres an hour.[2] It is true that
these permits were reserved for functionaries or

---

[1] Seneca, *Cons. ad Helviam*, 6.

[2] See *Histoire des mœurs romaines d'Auguste aux Antonins*, by M. Friedlaender, translated by M. Vogel into French. This excellent work, full of curious facts skilfully presented, contains an entire long chapter concerning journeys among the Romans. In it all those details will be found which I cannot give here.

couriers of state. It is rather surprising that it did not occur to this practical people, who so quickly seized the utility of things, to authorise private individuals to use the official post on payment, which would have rendered communication more rapid, and more closely bound the different parts of the Empire together. But authority was probably tenacious of its privileges, and stopped by a fear of diminishing its prerogatives. In the absence of the post, private persons furnished those who wished it with sufficiently convenient means of travel. At the gates of towns near the hostelries, which then as now bare a cock, an eagle, or a crane for their signs, and which endeavoured to attract passers-by with all kinds of engaging promises, it was easy to find carriages of every sort for hire, or to provide oneself with a horse or mule, by addressing those rich societies (*collegia jumentariorum*) which always had them at the disposal of the public. With these horses and these carriages one could go fast if one cared to do so. Suetonius informs us that Cæsar thus got over 100 *millia* (150 kilomètres) per day. But usually people were not in such a hurry; they went by short stages, lingering at good spots; they stopped when they were tired, and admired nature at their ease. A few years ago, this was still the way people used to travel in Italy. Some think there was none more pleasant, and regret that it has been given up

During the first century of the Empire there was no lack of reasons for travelling. Many of the people who were to be met with on the highways were functionaries on their way to rule distant provinces. Rome had conquered the world, and had to govern it. She sent her proconsuls and her prætors everywhere, and they took with them their lieutenants, their quæstors, their secretaries, their *apparitors*, their freedmen, and their slaves—a whole world, who were often on their way to live at the expense of the provincials. In the footsteps of the governor, and often in advance of him, travelled the farmers of the public tax, with their scribes and their agents, and those merchants who so well knew how to exploit a vanquished country. There were also, and in great number, students repairing to well-known professors in towns where learning flourished; invalids attracted by famous physicians, sulphurous waters, or healthy climates; devotees visiting, one after the other, all the important sanctuaries, and always with some question to put to renowned oracles; and then, people who had not found fortune at home, and were seeking it elsewhere. "All the wretches," says Seneca, "who hope to turn their beauty or their talents to account, stream into those great towns where virtue and vice are paid for more dearly than elsewhere." After those who travelled as a duty or from necessity, came those who travelled for pleasure. The taste for becoming acquainted with countries which contained fine monuments or recalled great memories, arose early. Greece first attracted the lettered, and thence they passed into the East. After Pharsalus, Cæsar does not fail to go and see the "fields where Troy was." Germanicus traverses Asia and Egypt, whose curiosities and hieroglyphics he makes the priests explain and read to him. It may be supposed that among these sincere admirers of the past, who piously visited its remains, there were persons who travelled for fashion and appearance, to do like all the world. There were some, too, as we know, who only undertook these long journeys in order not to remain at home. Great refined civilisations, which create so many wants in man, by habituating him to satisfy all his desires, and constantly stimulate the soul without contenting it, often bring with them a tiresome companion—*ennui!* "which," says Lucretius, "flows from the same source as pleasure," and suffices to render life unbearable. One always fancies that the best means of escaping it is change of place, and one hastens to leave one's home and one's country. In vain had the ancient philosophers repeated that we do not thus rid ourselves of our cares, that they faithfully follow us in all our excursions, and "ride on horseback behind us"; the philosophers convinced no one, and the *ennuyés* of the second century, like those of our own day, continued to seek everywhere for unknown sights and new pleasures capable of affording a moment's distraction.

Hadrian had all these reasons at once for running about the world. The most important and best of all of them was his desire to personally ascertain the state of the Empire. An administrator of his stamp was not unaware that it is good for the master to see everything with his own eyes. It was his custom to stop at the large towns which were on his line of route. He demanded an account of the manner in which they were governed, minutely studied their resources and their wants, and his passage was rarely unmarked by the construction of bridges, roads, and aqueducts which he had deemed necessary. Being also very fond of magnificence, after having busied himself with useful works, he did not neglect the monuments that serve for the decoration of a great country. He repaired the theatres and the basilicas; he had ancient temples rebuilt and raised new ones. So he always left the provinces full of admiration and gratitude. We have preserved the medals struck by them on the occasion of these imperial visits. They call Hadrian the restorer, the benefactor, the genius of the cities he had passed through, and decree him in advance the apotheosis he could not escape on his death. When he arrived on the frontiers of the Empire, he naturally redoubled his care and vigour. Nothing was forgotten. He saw that the fortresses, the dykes and the retrenchments were in good condition: he listened to the officers, consulted the engineers, inspected the legions, made them manœuvre before him, and, if he was satisfied with their evolutions, addressed to them one of those oratorical orders of the day of which so curious an example remains to us in the inscriptions of the third legion at Lambæsæ. But Hadrian did not travel solely in order to be useful to the Empire; he also thought of himself. This zealous administrator was at the same time a lover of art, a scholar, and a man of letters. When the town which he came to was one of those possessing fine monuments of the past, he liked to remain longer in it, showed it more kindness, and took occasion to return to it. His sojourn at Athens enraptured him: nowhere did he feel so happy, and there is no town so loaded by him with benefits, and where he built more monuments. His curiosity did not forget any of the spots recalling great memories. He, too, made his pilgrimage to Troy, and restored the tomb of Ajax, to whom he paid great honours. At Mantinea he went to see the tomb where Epaminondas rested, and composed for the Theban hero an inscription full of enthusiasm. In Egypt, he presided over the assembly of the learned in the museum, and took pleasure in embarrassing them by his captious questions. He also went to see the Pyramids, the Colossus of Memnon, and probably all the other wonders of the time of the Pharaohs as well. In these visits he did not think himself obliged to observe that cold and formal air which old Romans were careful to assume when from Rome, in order to appear more grave and dignified. He spoke the language of the nations whose guest he was, donned their dress, and did not disdain their usages. He doubtless thought that in order fully to enjoy a country and understand a people, one must enter into its manners and live like them. At Eleusis he had himself initiated into the mysteries; at Athens he presided over the feasts of Bacchus in the garb of an archon. This behaviour must have shocked people who held to the ancient usages. One of these malcontents, the poet Julius Florus, made some little malicious verses against the travelling prince, which must have been read with pleasure by all those who could not make up their minds to lose sight of the seven hills. "I would not be Cæsar," said he, "to hurry off to the Britons, and bear the snows of Scythia," etc. To which Hadrian replied in the same tone and in the same metre: "I would not be Florus, to walk about in the shops, rot in the taverns, and be eaten up by the gnats there:" and without any more care for opinion, he continued his roamings. He occasionally even made veritable innovations, and sought after spectacles which had hitherto been neglected. A poet of the first century who has left us an interesting description of Ætna, is much surprised at the indifference of his contemporaries for the sights of nature. "They pass over lands, they pass over seas," he says, "in order to visit great cities and fine monuments. They go to see famous pictures—a Venus whose tresses seem to wave like a river, or the children of Medea playing quite close to their cruel mother, or the Greeks who sadly surround Iphigenia and drag her to the altar, while a veil covers her father's countenance: they admire the statues which have made the glory of Myron and others, while they do not deign to look at the works of Nature, who is a much greater artist than they."[1] Hadrian deserves not this reproach. His passionate taste for the masterpieces of ancient art did not prevent him from being sensible of the great scenes of nature, and he is nearly the only person of his time of whom we are told that he travelled in order to contemplate them. He climbed Ætna, and we are still shown the ruins of an old house said to have been made there to receive him. He went up Mount Casius by night to see the rising of the sun, and was there witness of a terrible tempest. So he loved nature as much as he enjoyed the arts. This admiration of art and this love of nature will be seen again in the villa at Tibur.

---

[1] Lucilius, *Ætna*, 587.

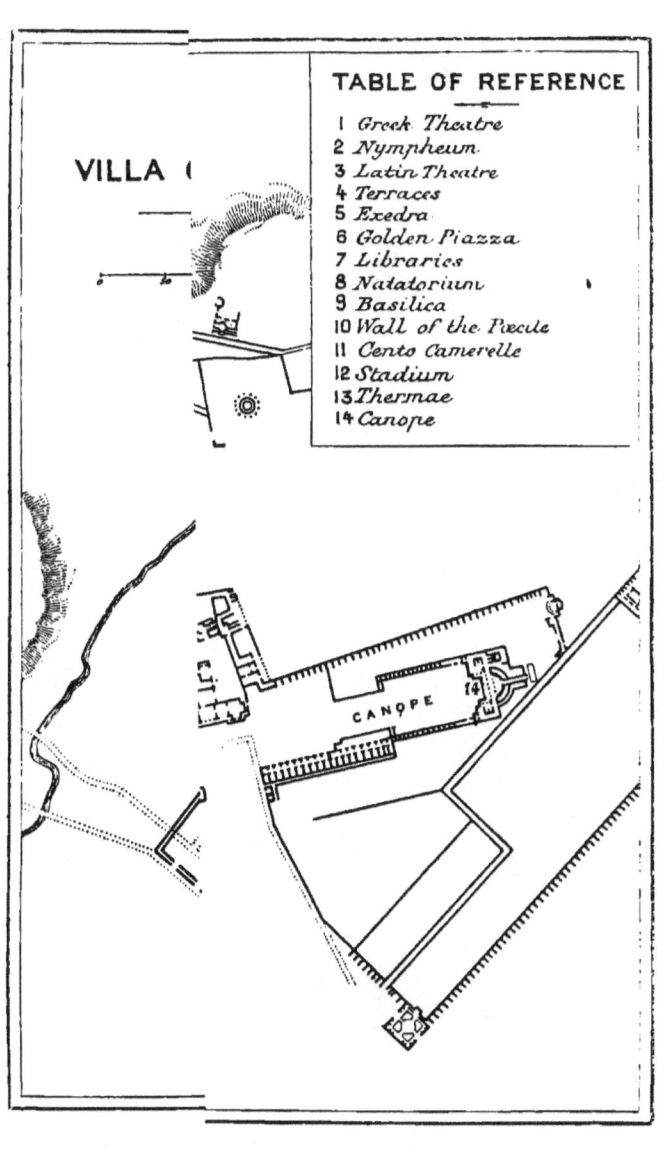

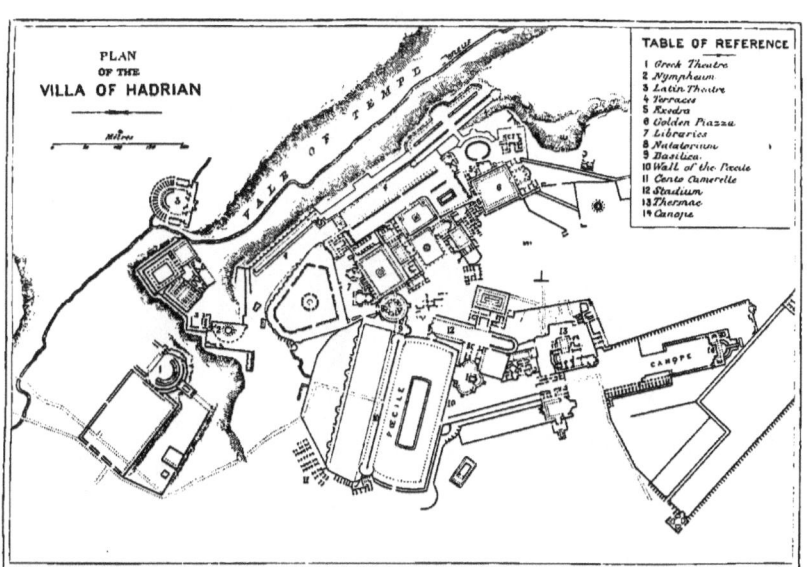

II.

SITE OF HADRIAN'S VILLA — MAGNIFICENCE OF CONSTRUCTION—THE EMPEROR'S PURPOSE IN BUILDING IT—PARTS WHICH CAN BE RECOGNISED — THE VALE OF TEMPE — THE PŒCILE — CANOPUS — THE PRIVATE DWELLING — THE *NATATORIUM* — THE RECEPTION APARTMENTS—THE *PIAZZA D'ORA*—THE BASILICA—THE THEATRES—THE LIBRARIES — THE PUBLIC LECTURE HALLS—HELL.

AGE put an end to all these wanderings. When Hadrian was nearly sixty, he felt the need of rest. Having no children, he began by choosing a successor. He first adopted Lucius Verus, who died before him, and subsequently honest Antonine. "Then," says an historian, "seeing that all was quiet, and that he might relax his cares without danger, he left the administration of Rome to his adopted son, and withdrew to his villa at Tibur. There, as is the custom of the rich and fortunate, he no longer busied himself with anything but buildings and feasts, statues and paintings. In a word, he had no further care but to pass his life in joy and pleasure." It must be concluded from this passage, that in 136, when Hadrian resolved to retire from affairs, the villa at Tibur already existed. When he began building it is not known, but it is certain that he passed the last three years of his life in beautifying it, finishing it, and putting it in that state of perfection which caused it to be considered one of his finest works.[1]

The site of the villa at Tibur is not only very pleasant, it is also extremely healthy; at that time the highest merit of a country house. Doubtless the plain of Rome, covered with trees and cornfields, filled with charming dwellings, with villas and gardens, did not resemble what it has become after several centuries of neglect: it was not yet a desert and a cemetery; yet, even at the time when it was richest and most peopled, the bad air was feared. Cicero highly congratulated Romulus on having succeeded in founding a healthy town in a pestilential country (in *pestilenti loco salubrem*).[2] We know that this pretended salubrity of Rome did not prevent the yearly heats from bringing fevers, and, as Horace puts it, causing wills to be opened. In the surrounding Campagna it must have been far worse. So, when one desired to build a villa, it was above all things necessary to choose its site well. That of Hadrian is situated near the last spurs of the Apennines, at the foot of the hill on which Tivoli is built. While freely open to the beneficent influence of the west wind, the heights surrounding it protect it from the *sirocco* and pesti-

---

[1] The map we give of Hadrian's villa is chiefly constructed according to Nibby, but, in the part near the Valley of Tempe, Nibby has often been corrected in accordance with M. Daumet's plan. It was not possible to mark on the map the result of the last excavations, which are not yet finished, and will doubtless involve some modifications in the manner of representing the imperial palace and the *natatorium*.

[2] Cic., *de Rep.*, II. 6.

lential breezes from the south. Two small parallel valleys run north and south, enclosing a plain rising in stages, and forming a kind of eminence three miles in length. It is on this plain that the villa was built. The ground contained many of those natural inequalities such as we preserve with care, and which seem to us the greatest charms of our gardens. The Romans, on the contrary, did not like them, and took great pains to level, by means of vast substructions, the soil on which their town or country houses were built. We also find these substructions, and in great numbers, in the villa of Tibur. Two little streams, descending from the mountains of the Sabina, traverse the two valleys, and join near the entrance of the villa to throw themselves into the Anio. Like almost all those of Southern Italy, they are nearly dried up during the summer,—that is to say, in the season when they are most wanted to be full. Their dearth was made good by aqueducts, whose remains have been found, which brought the fresh and wholesome water of the mountain in abundance, both to the dried-up beds of the rivulets and to the apartments of the palace.

What first strikes us on going over Hadrian's villa is its immense extent. Nibby affirms that it covered a surface of seven Roman miles. The Villa Braschi, which has been purchased by the Government, and is the only portion visited, does not include it all. If we advance towards the south, leaving the brambles, the dogs and the keepers, and cross the enclosures, we shall find other halls, more large and more beautiful, perhaps, than those shown to strangers. In order to join these apartments, so far apart and so like different quarters of a town, underground passages or *crypt porticoes* were dug, which allowed the prince to pass from one end of his palace to the other without fearing the heat or the importunate. In all these buildings marble was so lavishly used that the ground is still covered with it. In course of time it has crumbled up into a sort of dust, which glitters in the sun, and tires the eye with its reflections. The villa, when its building stood, must have been a marvel. It is impossible to cast one's eye over the restoration which M. Daumet has made of it without feeling dazzled by so much magnificence. It is difficult to imagine a richer and more varied assemblage of edifices. It is an incredible succession of porticoes, of peristyles, of buildings of all kinds and all dimensions. The domes of large halls, the round roofs of exedras are mixed with the triangular pediments of temples, while above the roofs rise high towers and terraces shaded by arbours. Some surprise is, however, mingled with our admiration. The *ensemble* of these vast edifices escape us. We admire their variety; we find a remarkable fecundity of invention and resource in them, but we are astonished at not seeing more symmetry. It is the impression also produced by the Forum, so full of temples, trophies and basilicas, and by the Palatine, with the five or six palaces that cover it. It will be remembered that we drew from it the conclusion that the Romans were less sensitive than ourselves to certain beauties which charm us, and that our large straight streets and regular squares would have probably left them somewhat cold. Hadrian's villa confirms this opinion. The architect seems to have added buildings one to the other as their want was felt, without troubling himself about the effect that might be produced by the whole. We must resign ourselves to the little taste shown by the Romans for symmetry. Let us reflect that, after all, it is not here a question of a palace situated in a capital, which must have a grand air, and give an advantageous idea of him who dwells in it, but a country house in which the architect is bound to consider convenience more than appearance.

Thus far, we have called all attention to nothing in Hadrian's villa that would not be met with in a lesser degree in others. There were no houses belonging to great personages that were not placed in healthy situations, provided, if necessary, with large subterranean works, richly furnished with running waters, adorned with precious marbles, or that did not contain an incredible number of magnificent apartments. The originality of the one with which we are busied consisted in the following. As nothing any longer interested Hadrian but his travels, he desired, even after he had given them up, to preserve living mementoes of them around him. His biographer relates that he attached to certain parts of the villa at Tibur the names of the most beautiful places he had visited. The Lyceum, the Academy, the Prytaneum, Canopus, the Pœcile, the Vale of Tempe, "and even," says Spartianus, "in order that nothing might be wanting, it occurred to him to make a reproduction of Hell there." This text may give rise to much discussion. There are authors who suppose that it ought to be taken literally, and who insist that Hadrian strictly and exactly copied what he had admired in his travels. Canina, especially, is most positive on the subject of these resemblances, and, if we were to believe him, there is not, in all these ruins, a fragment of wall but is the imitation of some important monument. He does not see that this is the way to make Hadrian very ridiculous. What effect could these reductions of mountains, these miniature valleys, and these monuments all heaped together, produce upon visitors? Hadrian, as we know, was a clever artist, a man of taste, a friend and enlightened admirer of Greek art—what pleasure could he find in tormenting nature to produce resemblances for him which could never be anything but incomplete? We are told that he wished his villa constantly to recall to him the marvels which he had seen; but these paltry counterfeits were rather calculated to spoil his recollections than to preserve them. Happily the text of Spartianus does not oblige us to admit all these exaggerations. He simply says that the Emperor so constructed his country house as to be able to inscribe in it the names of the most celebrated places he had visited (*ita ut in ea et provinciarum et locorum celeberrima nomina inscriberet*), which allows the supposition that he did not hold to very faithful imitations, and, for the most part, contented himself with an approximation. It was above all for the sites that great allowances had to be made. How could it be hoped to reproduce the wonders of nature in the little plain stretched out at the foot of Tibur? With regard to the monuments it was easier, and there were some, like the Pœcile, that could be pretty exactly imitated. Yet this exactitude was probably never pushed very far. M. Daumet bids us remark that in these Lyceums, these Gymnasiums, these Prytanea, that is to say, in these Greek monuments which the artist professed to copy, we everywhere find the Roman vault. "Is this not a proof," he adds, "that he did not pride himself on a scrupulous fidelity, and that, while keeping the foreign names of these buildings, he adapted them to the taste of his age and the usages of his country."

Of all those fine things which Spartianus has enumerated for us, many are impossible to distinguish now that all is in ruins. Yet there are others found again with almost certainty, and which help us to judge the rest—these are the Vale of Tempe, the Pœcile, and Canopus.

With regard to Tempe there can scarcely be a doubt. It would not be possible to place it elsewhere than in that kind of depression which separates the villa from the hills on which Tivoli rises. It was therefore situated towards the north-west side, along the little stream which archæologists call the Peneus. Certainly, there was neither Olympus, nor Pelion, nor Ossa here, nor those perpendicular rocks, of which Titus Livius speaks: "from whose summit the eyes and the soul are seized with a kind of giddiness,"[1] nor those dense woods "which the glance of men cannot pierce,"[2] and which gave to the real Vale of Tempe a mixture of grandeur and grace admired by all travellers. The grandeur is much diminished, but the grace remains. The little plain was in its nature not devoid of charm. They multiplied the shady places in it, they made it a

---

[1] Titus Livius, XLIV. 6. [2] Pliny, *N.H.*, IV. 8, 15.

place of pleasant walks, and as the alleys there were fresh and leafy, as it was delightful to rest there near the water, under the great trees, and recall the happy moments passed in roving through the beautiful Thessalian valley, they ventured to give it its name. On the side of the villa facing the plain large terraces extended, which are still to be recognised, with their porticoes and marble basins.[1] A vast exedra, supported on columns, with its back to the *Piazza d'oro*, commanded all the valley.[2] Thence one descended to the flower garden by gentle slopes. Only ruins remain of all this, but the site is still charming. Vigorous olive-trees grow in the interstices of the stones. When one sits down in the afternoon, under one of these great trees, with its knotty trunk and branches that take all kinds of strange forms, one has quite a carpet of verdure at one's feet, and, opposite, are the graceful spires of Tivoli, and the large modern villas with their arbours resting on stone pillars and resembling porticoes. It is difficult not to be struck with the beauty of the spectacle, and the valley seems so pleasant that one can easily pardon the fantastic Emperor for giving it so great a name.[3]

The Pœcile, on the other hand, looks towards the west, facing Rome. Turning to this side, we come to a large esplanade, where the inequalities of the ground have been corrected by considerable substructions. In order that nothing might be lost, the architect built,

---

[1] See No. 4 on plan.

[2] See No. 5 on plan.

[3] Moreover, let us not forget that the name had become general among the Romans, and that, in their villas, all fresh and pleasant valleys were called Tempe.

in the substructions themselves, several stories of apartments, commonly called the Hundred Rooms (*Cento Camerelle*).[1] Ligurio, who represented the Cæsars as he would the princes of his own time, fancied that they went nowhere without being followed by their soldiers, and supposed that these apartments were destined for the imperial guard, and other archæologists have accepted this opinion. In reality, the Roman Emperors, and especially those who were firmly established and had no sudden revolution to fear, did not drag armies after them, and as there were usually more slaves than soldiers in their country houses, it is natural to think that the Hundred Rooms, out of which it has been sought to make a Prætorian barracks, were simply the dwellings of the domestics. The esplanade, which extended above the substructions, was enclosed by an immense rectangular portico, in whose midst was a basin, of which some vestiges are still to be seen. One of the sides of the portico is preserved.[2] It is a brick wall 10 mètres high and 230 mètres long. Among so many accumulated ruins it has remained erect. When, after making our way with difficulty among those overturned blocks and scattered fragments of columns, we suddenly come face to face with this wall, so wonderfully intact, our surprise equals our admiration. We ask ourselves by what strange fortune it did not share the fate of the remainder, and how it was saved from the common ruin to which its very extent and height seemed the more to expose it? It can hardly be doubted that this portico is the one mentioned by Spartianus as the Pœcile, and which was

---

[1] See No. 11 on plan. [2] See No. 10 on plan.

the copy of an Athenian monument. The Pœcile of Athens, with which the description of Pausanias acquaints us, was chiefly famous on account of the paintings of Polygnotus. He had represented glorious deeds there, and especially the victory of Theseus over the Amazons, and the battle of Marathon. No trace of it now remains. Not knowing whether Hadrian imitated it faithfully, it is difficult to say how far the copy may give an exact idea of the model. It is however certain, that we can easily picture to ourselves what the Pœcile at Tibur must have been. On the two sides of the wall which has been so well preserved, rose columns of which only a few bases are left. They supported an elegant roof, and formed two porticoes communicating with each other by a door which still exists. This double portico was so devised that one of its faces was always in the shade when the other was in the sun, so that one could walk there in all seasons of the year and in all hours of the day. One had only to change sides, according to the hour, and could always find heat in winter and freshness in summer. The walls were probably covered with pictures, and these pictures must have been reproductions of those of Polygnotus. Time has destroyed them all, but it has not deprived this simple brick wall of its air of grandeur and majesty. It is certainly one of the finest Roman ruins remaining to us, and the admiration it inspires still further increases when we think of the Greek masterpiece which it recalls, and of which it is the last memento.

A little further on, continuing in the same direction, we come to a valley of somewhat small extent, and greater in length than in width, which archæologists, on the testimony of Spartianus, agree in calling *Canopus*. This name, like so many others, was given with reason. Upon a brick found in the valley, are read these words, which allow of no doubt: *Deliciæ Canopi*. We were just now at Athens, visiting the Pœcile, and now a fancy of the capricious Emperor suddenly transports us to Egypt.

We are to believe that Egypt was one of the countries which most struck Hadrian in his travels. That strange land, separated by its traditions, its customs, its language and its gods from the rest of the world, was not to be visited without the liveliest surprise. Since the Romans had become the masters of the universe, most of the nations had abandoned their laws and their usages in order to assume those of the vanquishers: Egypt, under all rules, remained faithful to her past. The Greek conquerors who came to reign over her, the prefects sent by Rome to govern her, changed nothing in her habits. Subdued for more than six centuries to strange dominations, she continued to live in her own way, built temples as in the time of Sesostris, and adorned them with hieroglyphs of which her conquerors understood nothing. This country, which resembles no other, and which Nature herself had made unique, became still more so by immobilising herself in her ancient civilisation. To contemplate this remnant of the past, so faithfully preserved, was to curious travellers a great attraction. So all those rich *ennuyés* who sought new sights and desired to escape awhile from the general uniformity, were happy to go over this corner of the world, which resembled nothing else. They did not fail to go and see the monuments of the Pharaohs, to look at the Pyramids, to hear Memnon salute the dawn, and to inscribe their names with thanks upon the pedestal or on the legs of the Colossus. On their return home, they asked sculptors or painters to reproduce what they had been admiring. Thus a false Egyptian taste got spread in the art of that epoch, which produced a few good works and many ridiculous imitations. From the great lords this taste descended to the other classes, and the citizens of Pompeii loved to paint on the walls of their houses improbable landscapes, with palm-trees, ibises, and crocodiles, in order to give some idea of that singular land to people who had never seen it.

Hadrian visited Egypt, like the rest, and it is not surprising that this curious and sagacious spirit should have been more struck with it than anybody. We have preserved a letter written by him from Alexandra to his brother-in-law, Servianus. In it the aspect of this great commercial city, where all the peoples of the East assembled, is grasped with great astuteness. He has, in particular, very waggishly described the activity of this people, busy in the hunt after fortune. "No one," he says, "is idle there. Some work glass, others make paper, others weave flax. Everybody has his work, and practises a calling. Even the blind, the gouty, and the halt find something to do. They have all of them but one god—money (*unus illis deus nummus est*)—'tis he alone that Christians, Jews, and all the rest adore." As happens in all great industrial cities where fortune is so shifting, people sought to enjoy quickly what might be so quickly lost, and gave themselves up to pleasnre with as much ardour as to business. The place of amusement of the Alexandrians, whither they went to distract themselves from their occupations and lighten themselves of their money, was the town of Canopus, situated five or six leagues from Alexandria. Canopus possessed a famous temple of Serapis, whither people proceeded from all parts of Egypt. Every evening the sanctuary was full of persons who came to ask the god to heal their own maladies or those of their friends. They slept in the temple after having offered fervent prayers, and during their slumbers they received in a dream the remedy that was to deliver them of their ills. But, more often, health was only a pretext, and folk went to Canopus as in our days they go to the thermal baths; less to be cured than to amuse themselves. The journey was made on a canal five leagues long, traversed continually by light barks, curved at the prow and at the poop, and having in the middle a sort of box, very like those of the gondolas of Venice.[1] The movement did not stop; day or night there resounded on the waters the love-songs of Egypt, famous throughout the world. On either side of the canal rose hostelries, abundantly furnished with all that could incite to joy and satisfy desires. People stopped there to drink the light wine

---

[1] Some are seen with this form in the famous mosaic at Palestrina, and a representative of one of those Egyptian feasts, which must have been so frequent along the canal of Canopus, is also found there. Under an arbour, covered with a fruit-laden vine, men and women are softly stretched, holding drinking goblets in their hands. One of the women raises the *rhytion* to her lips; a second points to the hanging grapes; others play the flute or stringed instruments, while around them flows the river, covered with lotus-flowers.

of Mareotis, which produced a gay and brief intoxication, and, the repast over, danced to the sound of flutes, under arbours or in the shade of trees. Thus they arrived at length at Canopus, where still more amusements awaited them than on the way. All was done there for pleasure, and it was impossible to imagine a more enchanting sojourn. "It was like a dream," says a contemporary author, "and one thought oneself transported into a new world."

Hadrian, who wished his villa to recall to him all the most striking things seen by him in his travels, took care not to forget Canopus. In accordance with his wont, he did not trouble to produce the Egyptian city exactly, which could not have been done in so small a space, and probably contented himself with a very distant resemblance. At the end of the valley, a sort of large niche or deep recess, ornamented with great magnificence, served, at the same time, for the temple and the water reservoir.[1] In the depression in the centre of the recess, the statue of Serapis, the great divinity of Canopus, must have been placed. In the lateral walls smaller niches contained other Egyptian gods. These statues are perhaps those found among the rubbish of the valley, and placed together in the Vatican museum. From all corners of the building water flowed in abundance. It descended the marble steps, or rebounded on superposed vessels, and thence fell into a large semi-circular basin. A sort of bridge or passage, placed over the basin and adorned with columns which sustained the roof,

---

[1] See No. 14 on plan.

enables visitors to pass from one side to the other, and to see the cascades more nearly. The water flowed beneath, and fell into a canal occupying all the middle of the valley. This canal, hollowed out of the tufa, was 220 mètres long by 80 mètres broad. Elegant barques, doubtless made on the model of those of Alexandria, were reserved for the Emperor and his friends, and on the quay the remains of the steps are still seen, to which the boats came to fetch them when they wished to amuse themselves on the water. On one side of it the ruins of a score of two-storied halls have been found, sheltered by a fine portico. This was perhaps an imitation of the voluptuous hostelries where the traveller bound for Canopus was so happy to tarry. Those of Hadrian's villa probably did their best to deserve the fame acquired by the others. What passed there may be guessed, when we remember that Hadrian was passionately fond of pleasure, and never took the trouble to conceal the fact. Marcus Aurelius made some allusions to these corrupting spectacles, when, later on, he recalled the dangers that had threatened his virtue in his youth, and thanked the gods "for having cured him of the passion of love to which he had for a moment yielded."

Of the parts of the imperial villa enumerated by Spartianus, the above are those still found, and which may be pointed out with fair accuracy. Thus, we still possess, and can still go over, what the capricious Emperor called his Vale of Tempe, his portico of the Poecile, and his "delights" of Canopus. This is something, but it is possible to proceed further without compromising ourselves. This immense mass of ruins must have contained apartments which the Emperor was obliged to construct, which the exigencies of his position solicited for his comfort or his pleasures, which his wants or his tastes rendered necessary for him, and it is not unreasonable to hope to find them.

And, first of all, it cannot be doubted that he reserved a portion of this vast palace for the uses of his private life. An aged and invalid prince, so carefully building an asylum for his last days, must above all things think of his pleasure and his comfort. But where are we to locate his private dwelling? From the time of Ligorio the ruins extending to the west, along the Vale of Tempe, have been called the *Palazzo imperiale*. M. Daumet thought himself obliged to place it elsewhere. He recalled that, in the villas in which opulent Romans sought shelter during the summer heats, as well as in those still remaining from the Italian Renaissance, the dwelling is always situated above accessory buildings, in the most elevated part of the grounds. It is natural, indeed, that the master should wish to command the plain and enjoy the most extensive and varied view possible. If it was thus in the case of Hadrian's villa, we must look for the prince's private abode a little further to the south, on the plateau, where Ligorio thinks he finds the Academy, and Canina the Gymnasium, nor does M. Daumet hesitate to place it in this spot. However, the excavations made a short time since, show him to be wrong. In digging in the place indicated by Ligorio, chambers of moderate extent have been discovered, with corridors and porticoes whose proportions recall those of the fine houses of Pompeii. It is certainly a dwelling appropriated to everyday life, and as it is, nevertheless, sumptuous and quite close to the great reception apartments, we are free to think that the Emperor built it for himself. So Ligorio was probably not mistaken in placing the *Palazzo imperiale*, that is to say, the prince's private habitation, close to Tempe.

Near the chamber where he slept at night, a Roman or a Greek thought nothing more necessary to his existence than a bathroom. So in the villa at Tibur, the construction of *nymphea* and *thermæ* was not neglected.[1] They were wanted for the prince, for his friends, and for his servants. It is doubtless to this use that a circular building, situated between the private apartments and the Pœcile, was destined, and which is perhaps the most curious and the richest thing found in the villa.[2] The foundations are well enough preserved to enable us to reconstruct the plan without much difficulty. A circular portico, supported on columns of *giallo antico*, of which some remains strew the ground, surrounds one of these little streams called by the ancients *euripes*. The canal, lined throughout with white marble, in which the water was to flow, is about 5 mètres broad and a little over 1 mètre deep. The space enclosed by the little stream forms a kind of islet, which marble bridges join to the exterior portico.

---

[1] An elegant *nympheum* and very vast *thermæ* are thought to have been discovered in the villa. See Nos. 2 and 13 on plan.
[2] See No. 8 on plan.

R

By a not ungraceful caprice, in the middle of the round island there is a species of square court, which was doubtless meant to be ornamented with some statue. Little rounded chambers, niches opening on the *euripes*, whence fountains flowed, occupied the unequal segments between the rectangular line of the court and the curved line of the canal. Nothing is more original or more pleasing to the eye than all these ingenious combinations. The floor of the chambers, of the court, and of the porticoes is covered with broken marble. Numerous remains of columns have been found here, with fragments of bas-reliefs, representing marine monsters—Tritons, Nerides, and little Cupids mounted on hippocampi. What could have been the destination of this fine edifice, constructed with so much care and refinement? The most probable opinion seems to be that of Nibby, who calls it a *natatorium*, and makes a kind of piscina of it. The small chambers surrounding the *euripes* were perhaps cabinets for repose, or might serve the bathers to undress in. The remains of steps have been found leading from them into the canal. The nearness to the imperial palace, and the magnificent decorations of these baths incline one to the belief that the Emperor had reserved them for himself, and, indeed, they are quite worthy of this voluptuary, this friend of refined pleasures. It is difficult to imagine a spot where one might repose more pleasantly in the oppressive summer hours than in these elegant halls, in the midst of all the riches of a refined art, beside this *euripes* noiselessly circling in its marble bed, and lulled by the murmuring of the water softly falling from the fountains.

Not far from the habitation of the prince were the reception apartments. We are bound to believe that, although Hadrian, in building his villa, paraded his taste for retirement, he did not abandon his duties as Emperor until the end. However numerous the friends of kings may be, it was not for friends alone that these immense halls, which still excite our astonishment, were built. There are some magnificent ruins near the *Palazzo imperiale*, along the Vale of Tempe. This is the part specially studied by M. Daumet, who has tried to restore them for us to nearly the same condition they were in at the time of Hadrian's death. In order to get to the chief halls, a long series of various edifices had to be passed through, which must have greatly impressed visitors. An octagonal vestibule led into one of those courts, called by the Romans peristyles. There were many in the villa, but this one must have been more spacious and more beautiful than the others. So many rich remains of it have been found, that the architects by whom it was cleared named it the *Piazza d'oro*.[1] It was surrounded by a portico, with columns of *cipollino* and Oriental granite. A pavement of pink marble covered its floor, and statues, whose bases are believed to have been found, completed this magnificent decoration. At the end of the peristyle, facing the vestibule, rose a vast hall, terminated by a semicircular recess. At the four corners of the hall were niches lighted from above. M. Daumet thinks they were made to contain statues, and the care taken to light them well induces the belief that they must have been the works

---

[1] See No. 6 on plan.

of renowned artists. It is known that this favourable arrangement, which allows of the better enjoyment of works of art, has been reproduced in the Belvedere Court of the Vatican. So much magnificence seems, indeed, to show that this fine hall, with the peristyle preceding it, was reserved for imperial audiences, and that it was here that the prince admitted to his presence the envoys of towns and provinces who came to see him. We may also connect with these official apartments, where Hadrian fulfilled his imperial function, a hall in pretty good preservation, which we pass through on our way from the *natatorium* to the Pœcile.[1] It has been styled, in turn, a temple and a place of meeting for philosophers (*schola stoicorum*). Hadrian did not like philosophers, and least of all the Stoics, enough to build them so fine an edifice. I should be rather tempted to see in it a basilica, for it resembles the one found on the Palatine. We know that Trajan was accustomed to assemble in his villa of the Hundred Chambers (*Centum Cellæ*) a sort of privy council, composed of senators and magistrates, to judge with him causes whose decision he had reserved to himself. These were usually delicate matters concerning officers of his army or persons of his household. By day, advocates were heard and sentences pronounced; in the evening the Emperor admitted the judges to his table, and the repast ended, they indulged in pleasant conversation or listened to mimes and comedians.[2] If Hadrian followed Trajan's example, which is likely

---

[1] See No. 9 on plan.
[2] Pliny, *Epist.*, VI. 31.

enough (for he was a great disciplinarian), and assembled this kind of tribunal in his villa, they probably held their sittings here.

Lastly, let us not forget that Hadrian was not only a blameless Emperor, who made a point of fulfilling exactly the duties of his position, but that he was also a very refined scholar, felt a great attraction for intellectual pleasures, and loved much to imitate the Greeks. We must, indeed, believe that these tastes of the old prince left some traces in the villa built by him. Near the Pœcile a stadium, with very considerable dependencies, has been found in pretty good preservation.[1] All the Emperors who loved Greece affected a passion for the games of athletes, much as, in the last century, great French lords, who wished to follow the fashion of the English aristocracy, never spoke of anything but horses and jockeys. Scenic displays were still better provided for, and in the villa there are at least three theatres. One seems to have been an odeum, another, the best preserved of all, situated at the point where the villa is now entered, is preceded by a large square piazza, which must have served as a promenade for the spectators. Certain details of construction have led to the belief that it was a Greek theatre.[2] The Latin theatre is a little higher, on the same side as the Vale of Tempe.[3] It is now much deteriorated, but it is said that in the last century the marble linings of the orchestra, and the bases of the statues which ornamented the *podium*, were still to be seen. It must be owned

---

[1] See No. 12 on plan. [2] See No. 1 on plan.
[3] See No. 3 on plan.

that this abundance of theatres, in an age when dramatic art was so little cultivated, is rather surprising. The existence of the Greek theatre, could, indeed, be understood; a lettered prince like Hadrian, endowed with a taste for refined things, might like to listen to the pieces of Menander there. This great poet, who knew life so well, and had so delicately portrayed it, kept all his empire over a delicate and distinguished society. He was studied in the schools, read in the world, and we know that at Naples, in the first century, he was played. But what could have been represented in the Latin theatre of the villa at Tibur? Is it likely that they went back to Plautus, Cæcilius, and Terence? These revivals of admiration were then much in fashion. Hadrian piqued himself on preferring Ennius to Virgil, and Frono, in his correspondence, speaks on every occasion of the old *atelanæ;* but to admire ancient writers in one's study, or cite fragments from them in one's writings, is not the same thing as to produce them on the stage before people who understand them only with difficulty. Perhaps the Emperor, in order to seem to protect letters, welcomed in his country theatre the rare works still composed by a few men of talent. They were generally rather poor imitations of the Greek dramatists, made for fashionable drawing-rooms, and which could scarcely succeed before a real public. Perhaps, too, Hadrian, who towards the end of his life was morose and sought distraction, had actors of popular pieces sent to his villa to play before him: two mimes were then in favour with the Roman populace; one representing a robber-chief at loggerheads with justice, and laughing at the people who try to take him;

the other in which a lover, surprised by the unexpected return of the husband, is obliged to hide in a box; two subjects which from that time have not ceased to amuse the mob, and sometimes clever people as well.

There is no doubt that there were libraries in Hadrian's villa—probably a Greek and a Latin one. They are believed to have been recognised in two buildings placed close to each other, and containing several rooms.[1] The only reason for thinking so is, that they are situated according to the rules of Vitruvius, who holds that books should receive the morning light. Above one of these buildings rose a three-storied tower, which may have served as an observatory to a prince fond of astrology. According to usage, these libraries must have contained the busts of great writers, as well as their works. A certain number have been found in the neighbourhood of Tivoli, of which one, at least, comes from Hadrian's villa. Each of them bears a short inscription, characterising the personage whose features it produces. Below the sage Solon are read these words:—"Nought in excess." The prudent Pittacus teaches us that "The opportunity must be seized," and melancholy Bias that "The great majority of men are wicked."[2] This custom of decorating libraries with the portraits of the great men whose works they hold, existed in the time of Cicero. Sad, discouraged, and foreseeing the end of the Republic when he beheld

---

[1] See No. 7 on plan.
[2] These *hermes* are now placed in the Hall of the Muses of the Vatican museum.

the dishonest attain to the highest honours, he took refuge in study, lived in the midst of books, and wrote to his friend Atticus: "I love better to be seated at your house, on this little bench, which is beneath the image of Aristotle, than in their curule chairs."[1]

Neither do I hesitate to believe that the villa at Tibur must also have possessed a hall for public readings. Hadrian was very fond of them. At Rome he had had the Athenæum constructed, whither rhetoricians and poets came to recite their writings. He probably did not neglect to provide his country house, where he had more leisure, and could listen at his ease to the authors whom he loved, with some building of this kind. Unfortunately, it has not yet been possible to find it in the midst of all these ruins, any more than the Lyceum and the Academy. Perhaps the little theatre of which a few remains were found at the end of the villa, and which archæologists call an *odeon*, was destined to this use. According to Hesychius, the odeon was reserved for the displays of the rhapsodists and players on the cithern.[2] It was natural that it should also be used for public readings, and it may indeed be concluded from a curious passage in Horace, that people really met in the theatres to listen to the works of authors of renown. He tells Mæcenas, in order to make him understand whence come the enmities which pursue him, that he has not been pardoned for having refused to read his works in public. At the very moment when Pollion has just instituted these literary feasts, and when all Rome, not knowing

---

[1] Cicero, *Ad. Att.*, IV. 10.   [2] Hesch, s. v.

what to do with its leisure, is rushing to them, he seems to condemn them, by abstaining from taking part in them. His only reason is that it is repugnant to him to make a spectacle of himself, "in a theatre," for the crowded throng:—

> "*Spissis indigna theatris*
> *Scripta pudet recitare:*"[1]

But others had not the same scruples. Ovid likes to recall that in his youth he read his love verses "to the people,"[2] and we are told that Statius, when he consented to promise that on a fixed day he would read his poem, made "the town" happy.[3] Although allowances must be made for the exaggerations of poets, "the people" and "the town" mean very numerous assemblies, which could not be held in ordinary rooms, and it is probable that the full theatres (*spissa theatre*) spoken of by Horace, are also meant in these cases. Even when the readings drew fewer people, and the meetings were held in more modest premises, if these were no longer real theatres, they must, at least, have had their form. Juvenal greatly pities the poor authors who, in order to make themselves known, borrow an old, unused room from some great lord, and furnish it at their own expense. One sees from the terms he employs, that they arrange it so that there may be an orchestra and tiers of seats,—that is to say, what

---

[1] Hor., *Ep.* I. 19, 41.
[2] Ovid, *Trist.*, IV. 10, 55.
[3] Juvenal, VII. 83: *Lætam cum fecit Statius urbem.* And, further on, he not only calls the audience *populus*, like Ovid, but *vulgus*.

really characterises a theatre.[1] The orchestra, whence one sees better and hears more closely, is reserved for personages of importance. It must be furnished with commodious seats, so that being more at their ease, they may be the more disposed to admire. On the benches raised one behind the other, crowd obscure friends, clients, dependants, all those whom one invites that they may swell the number and applaud: these are the noisy portion of the audience. The great personages in the orchestra only give vent to a little murmur when they are pleased, but the friends in the farthest seats show their approbation by shouting out and stamping. Opposite, in a sort of raised tribune, towers the reader. He comes and takes his seat there with a modest air, "well combed," says Persius, "draped in his new toga, wearing rings on his fingers, his larynx made supple by an emollient potion, and gazing on the audience with a caressing eye."[2] If he reads agreeably, if he has chosen his audience well, if he possesses some resolute friends in the orchestra and vigorous clients on the benches, his first words will be received with favour, murmurs of approval will soon change into applause, and, as happens in these well-prepared assemblies, the audience, exciting each other, will ere long rise to transports of enthusiasm.

It is thus that so many mistakes arose at this epoch as to the real merits of works, and agreeable frivolous productions, whose success ought not to have lasted beyond the day, were saluted as marvels destined to live for ever. It would be very interesting to come upon one

---

[1] Juvenal, VII. 46.   [2] Persius, I. 18.

of the halls in which these little scenes took place. I do not know whether we shall be so fortunate as to find some remains of them in the villa at Tibur. In any case we may be sure that it will resemble this odeon of which I spoke just now, and that it will always be some kind of small theatre.[1]

In conclusion, it only remains for us to allude to Hell; for in the Tibertine villa there was also a reproduction of the infernal regions. Hadrian, his biographer tell us, put them there in order that nothing might be lacking. Archæologists have endeavoured to find its site again, but it will be difficult to succeed so long as we do not know on what model the Emperor built it. Was it a work of individual fancy, or did he adhere to the descriptions in the sixth book of the *Æneid?* We do not know. The curious thing is, that it should have occurred to him to place Tartarus and Elysium in his country house. Does it not prove that his contemporaries were beginning to busy themselves strangely with the other

---

[1] In the month of March 1874, while excavating on the Aventine, at the spot where it is believed the gardens of Mæcenas were situated, a vast hall was found, magnificently decorated, and forming at one of its ends a hemicycle, around which seven concentric rows of benches rise to the ceiling like an amphitheatre, while at the other extremity they came upon what were thought to be the traces of a sort of tribune. This reading-hall had been discovered, and it was called the *auditorium Mæcenatis*, under which name it is still known. (See *Bull. d'arch. munic.*, 1876, p. 166, etc.) But some doubts have since arisen with regard to this attributed designation. M. Mau, in the *Bulletin de l'Institut de correspondance archéologique*, 1875, p. 89, maintained that it was only a kind of hot-house, and that the benches had served to put pots of flowers upon.

life?[1] As for himself, I do not believe that he was much tormented by the thought of it. This wary politician and sceptical wit was not of those on whom the mystic religion of the East, and the new feelings they spread in the world, could have had much hold. We are told that, when he felt death coming, he was sufficiently master of himself to compose little frivolous verses, in which, addressing his little, trembling, charming soul, he tells it, with an accumulation of strange diminutives which cannot be rendered, "Thou goest to the regions pallid, stern, and bare, where thou canst no more yield thee to thy wonted sports." How had he represented these "regions pallid, stern, and bare" in his villa? We must make up our minds to remain in ignorance of this.

III.

DID THE ROMANS UNDERSTAND AND LOVE NATURE?— THE REASONS THEY HAD FOR LEAVING THE TOWN— HORACE AT TIBUR—LIKING OF EVERYBODY FOR THE COUNTRY—HOW PLINY THE YOUNGER LIVED THERE —HIS VILLAS — HIS GARDENS — SITES PREFERRED BY THE ANCIENTS—THE VIEW FROM THE PŒCILE.

THE above description of Hadrian's villa explains that it has sometimes been severely judged. Certainly nothing less resembles a country house as we now

---

[1] The day when Caligula was killed, he gave games to the people in which Egyptians and Ethiopians represented scenes in the infernal regions. The spectacle was to take place in the evening and be prolonged into the night.

understand the term. This profusion of buildings, this crowding of edifices, this stadium, these theatres, this Lyceum, this Academy, are bewildering to our habits of thought. There is nothing rustic here, nothing smacking of the fields; all seems to be showy, worldly, and got up. Perhaps we should merely conclude from this, that the Romans understood the country differently from us. But people go farther still, resolutely affirming that they did not like it at all, and the villa at Tibur serves as an argument for those who would prove that they never had either the understanding or the taste for nature.

This is a reproach very generally made against the Romans, and in our eyes it is a serious one. We unanimously profess to be passionately fond of nature; it is more than ever good form to go and visit famous ruins, and we should all feel extremely hurt if accused of not properly admiring them. Nobody would be found among us with the courage to say, like Socrates: "Not only do I not leave my country, but I never put my foot outside of Athens, for I love to learn—and the trees and the fields will teach me nothing."[1] This is a confession that would make us blush. The fields and the trees have now grown more obliging, and there is no one, even among the most simple and cockneyfied, who does not profess to gain something by communing with nature. The curious have noticed the epoch when this taste for natural beauties became so lively. It was born in the middle of the eighteenth century. Rosseau first brought

---

[1] In the *Phædra* of Plato.

the mountains into fashion, and the glaciers were found out by following his footsteps. Since then Switzerland, which used to be considered a barbarous country, has become the obligatory pilgrimage of all persons endowed with self-respect. It is talked about every day, read about everywhere, and makes us very proud of ourselves. I do not mean to say that people are altogether wrong. Certainly, for a century past, the feeling for nature has become broader and more general; but, on the other hand, one should not exaggerate, and pretend that it was lacking in the Romans. They understood and loved it after their fashion, and I do not think it useless, since the occasion offers, to inquire what was their particular manner of loving and understanding it.

The Romans had come from the fields, and the country was long their favourite sojourn; but, later on, the town drew them to itself, and very few resisted the attraction which they felt for it. Great personages who aspired to public functions were, indeed, forced to settle there, in order to be always before the eyes of their electors. They were followed thither by the small proprietors of the Roman Campagna, when want obliged them to sell their fields to their invading neighbours. Then, after the others, came the free labourers, who could no longer find employment except in painful or dangerous works, in which the rich man feared to risk his slave. These poor people ended by wearying of the rude existence they were condemned to, and knowing that in town they would be amused and fed at the expense of the treasury, they hastened to emigrate thither. When they had once received their *tepera* of corn or oil in the public distributions, or their *sportula* at the doors of the rich; when they had once acquired the habit of assisting at those spectacles of every kind which filled a third of the year, there was no longer any means of sending them back to the fields again. Sensible people were indignant at seeing a population of idlers continually growing, from whom in moments of public danger not a soldier could be drawn. Varro eloquently complains that the country had become depopulated since the husbandmen have slipped into the town, one after the other, and that "those strong hands which used to work the soil are now only busied in applauding at the theatre or the circus." But these honest complaints were not listened to, and the impetus once given, it could not be stopped. From the time of Augustus, the great city had absorbed the inhabitants of the surrounding country. The country was now only filled with vast pastures and country seats, while the old towns of Latium or the Sabina, that had for so long stemmed the fortune of Rome, fell into ruin.

The sojourn at Rome must assuredly have been very pleasant. Distractions and pleasures of all kinds were found there in abundance, suited to all tastes and all fortunes. Yet it could not escape the ordinary condition of large towns. The ardent life led in them famishes by producing an unbearable weariness. The perpetual tension to which the mind is doomed exhausts it, the noise stuns, the whirl of business one is thrown into makes one giddy, and one can hardly bear this general agitation, the sight of which at first pleased the eye. And just as warmly as we desired to be taken out of ourselves by exterior movements, do we wish to become our own again, and belong to ourselves for a moment. The most frivolous and worldly people feel strange yearnings for solitude and quiet, and endeavour to satisfy them. Milton has described, in beautiful verses, the joy of one of these prisoners, who, one summer morning, shakes off his chains and flies to the fields. Never did the meadows seem to him so green and the sky so pure. He listens to all the rural sounds, he inhales with joy the odour of the mown grass; he enjoys that broad and pure horizon which rests the eyes, and that warm, sweet air which expands the heart. All strikes and charms him, and the sights he has seen a hundred times appear to him new. There he is sensible to beauties he had never perceived, although he always had them before his eyes—he has discovered the country. I imagine that these were also the impressions of many Romans, who one day found courage to break their bonds, and went to ask the fields for a little rest and calm, and that it was thus that the weariness of worldly enjoyments that produced in them the taste for the pleasure of the country.

The poet Horace was, I believe, of this number. Nobody has celebrated the country more than he. From the manner in which he speaks of it, it would appear as if he were made to be happy only there, and had never loved anything else. One feels, however, that this taste was not so natural in him as in his great predecessor Lucretius, and his friend Virgil.[1]

---

[1] This is what seems to be noticeable in the landscapes which he loves to draw. Whatever their merit, they have always something less deep and more mundane than those of the other two poets.

Rome suited him very well during the first years. He found sights there which excited his inquisitive mind and stimulated his satirical spirit. It seemed very pleasant to him there as long as he could walk alone from the Forum to the Field of Mars, and look freely at the tumblers and fortune-tellers, but when the friendship of Mæcenas had turned him into a personage, and he could no longer leave his house without being beset by strangers who congratulated him on his good fortune, bores who questioned him on public affairs, and suitors who craved his help, the town became abhorrent to him. These importunities grew so hateful, that he nearly lost his usual moderation over them. He desired retirement with a passion which cannot but surprise us in a sage who professed to wish for nothing with too much ardour. So he lived very happily in his little country house. But I am tempted to believe that what lent the greatest intensity to his bliss was the recollection of the importunities of the town he had left. Perhaps he would not have thought that he made "repasts of the gods" there, had he not recalled, while seated unceremoniously at table with a few neighbours, the tedium of the great dinners at Rome, with their tyrannical laws that obliged one to drink as many cups as the king of the feast chose, and their unbearable conversations, whose sole staple were recent scandals and famous actors.

Mythology holds a great place in them, and it is not always, as with Lucretius and Virgil, the simple rendering and sincere expression of the great phenomena of nature. It is often merely one of those processes used by a clever man to throw charm into his descriptions.

The mischievous have remarked that he never seems so taken with the country as when he is kept in town. One day, at Rome, when he has suffered all kinds of solicitations and worries, he gives vent to the following exclamation, into which he has put all his soul: "*O rus quando ego te aspiciam?*" He appears to cool down towards his cottage when he gets there, and often wishes to leave it after having been in it but a few weeks. This is an inconstancy of which he humbly accuses himself, but has great difficulty in correcting. "Lighter than the wind," he says, "I wish to be at Tibur when I am at Rome, and I regret Rome when I am at Tibur!" Here, indeed, is the impenitent worldling, who thought himself cured because he felt a moment's disgust for those pleasures that enchant him, and who does not delay to resume his old yoke as soon as his ill-humour is past. It is only towards his life's end that his conversion became complete. He then got to love the country much more than his best friend would have had him do. He even broke his word to Mæcenas on this account, and, after having promised to be away only a few days, kept him waiting for whole months.

Horace's case must have been that of many Romans of his time. There were not a few, who, like him, became very fond of the country because they had been too fond of the town. These contrasts and revulsions are not rare with people who go to extremes in everything. When weariness and *ennui* drove them from Rome, they began by wandering around the great city, which they hardly dared lose sight of. They chose to separate themselves from it as little as possible, and built themselves villas quite close to the gates, along the high roads on the two banks of the Tiber. But they soon discovered that these villas and gardens, that cost so dear, did not give them relief. The town they would fain have flown from came and found them out there. The poor always in their way follow the example of the rich. Rome was oppressive to them as well, and they did not wish to stay in it. On holidays, the whole population of the poor rushed into the taverns of the outskirts, along the river, into the sacred woods, about the temples. They danced, "each with his female other," says Ovid,[1] and they dined in the open air, or under tents of foliage. It was a noisy, inconvenient neighbourhood, and it was scarcely easier to be quiet in the vicinity of Rome than in Rome itself. So they were obliged to go farther—to Tusculum, to Præneste, or to Tibur; and when these spots near the town, becoming fashionable, were in their turn too much frequented, and the calm and retirement sought for were no longer to be found there, they had to go further still. It is thus that all Italy, from the gulf of Baiæ to the foot of the Alps, became covered with elegant villas. "When," said Seneca to the rich men of his time, "will you cease to choose that there shall be no lake undominated by your country houses, no river unbordered by your sumptuous buildings? In every place where hot waters spring, you hasten to erect new asylums for your pleasures; in every spot where the shore forms a curve, you raise some palace, and not content-

---

[1] Ovid, *Fastes*, III. 525.

ing yourselves with the firm land, you throw dams out into the waves, in order that the sea may be included in your constructions. There is no part of the country where your dwellings are not seen to shine, sometimes on the tops of hills, whence the eye roves over vast extents of land and sea, sometimes raised in the midst of the plain, but at such heights that the house appears a mountain."[1]

It was not the rich alone who felt a craving to fly the town and breathe the air of the fields. Well-to-do freedmen, small citizens, and, above all, men of letters, more in love with silence and liberty even than the rest, were glad to possess, somewhere far from the crowd and the noise, what Juvenal calls "a lizard hole." Suetonius, whom his erudite works had not enriched, one day took it into his head to buy a small domain and not pay too dear for it. At his request, Pliny, who protected him, charged an important personage to mediate in the business. "What tempts our friend," he told him, "is the neighbourhood of Rome, the facility of the communications, the simplicity of the buildings, the small extent of the domain—large enough to distract, yet too small to engross him. It suffices studious men like himself to have enough ground before them to repose the mind and rejoice the eyes. They want scarcely more than a small bordered path, an alley to lounge in, a vineyard with whose every vine they are acquainted, and a few trees whose number is known to them."[2] Is this not still a real scholar's garden of our own day?

---

[1] Seneca, *Epist.*, 89, 21.   [2] Pliny, *Epist.*, I. 24.

Among these lovers of the country of all ranks and conditions who hastened to fly the town at their first leisure, there were indeed some who, like Horace, soon repented of having left it. Solitude bored more speedily than noise had tired them. They could not overcome their regret for the pleasures of the world. Could they long remain far from the games of the circus or the amphitheatre? "It was indeed indispensable," says Seneca, "to see a little human blood flow,"[1] and they hastened to return to Rome faster than they had left it. But this was the exception: generally rich Romans remained in their villas as long as they could. They had some on the tops of mountains, and on the borders of rivers for the summer season, and others, sheltered from the sharp winds, to live in during the winter. Some were very far away from Rome, and people went to them in the long holidays,—for instance, in the autumn, during the vintage vacations. They repaired to those quite close to the town, only when there were but two or three days of leisure at disposal. Thus, they dwelt in Rome solely when absolutely detained there by business, and sought the country even in town. "The populace," says Pliny, "were quite content to put flowers in their windows:"[2] poor flowers which must have had great difficulty to live, without air and without sun, in the narrow streets of the ancient city. Those who could have a house built for them took care to reserve room behind the *atrium* for a little garden with a few trees, which they called a grove, a little thread of water in a marble river bed,

---

[1] Seneca, *De tranq. animi*, II. 13. [2] Pliny, *Nat. Hist.*, XIX. 4, 19.

which they called *curipus,* and at the bottom a grotto of rock-work, beside a flying perspective of trees painted on the wall, so anxious were they to deceive themselves and forget that they were in the midst of a large town.

Here we have a community which appears indeed very much in love with the country, yet let us not forget that its taste for the fields was chiefly born of disgust for the town; for many signs show this. It is easy, I think, to see that the people who lived in those large villas were rather men of the world who wished to recuperate, than disinterested friends of nature. They did not go into it only to live in a sort of mute contemplation of rural beauties, and they would have been thought in the wrong had they shut themselves up in it to leave it no more. In the time of Tiberius, Servilius Vatia, an important personage of Rome, doubtless alarmed and disgusted by all he had seen in the Senate, caused a magnificent villa to be built for him near Cumea, and passed his life there. It does not occur to us to blame him for having retired from so much peril and shame, and no one will think of pitying him for having lived in such a charming country, yet the Romans found it extremely difficult to understand how, even under the Empire, one could voluntarily exile oneself from society and from public business. Servilius Vatia appeared to them to have buried himself alive, and Seneca tells us that every time he passed near the beautiful villa at Cumea, he could not help saying: " Here rests Vatia."[1] The masters of these country houses, then, were usually people engaged in the activity of business and the

---

[1] Seneca, *Epist.*, 55, 4.

movements of life — financiers and politicians who sought repose from old fatigues and strength for new ones, and writers who came to rest their minds and refresh their imaginations in solitude. "Here," says Pliny, quite happy to have reached his house at Laurentum—"here I hear no troublesome noises, here I commune only with myself and with my books. O sea, O shores, my true study rooms, what ideas you cause to arise in me, how many works you dictate to me!"[1] Being very fond of talking to us of himself, he draws for us the picture of the life he leads there, hour for hour. "I wake up when I can, usually towards the first hour (six in the morning). My windows at first remain shut, for I have remarked that silence and darkness stimulate the mind. If I have some work begun, I busy myself with it. I arrange all, the ideas and even the style, as if I were writing and correcting. I work thus, sometimes more and sometimes less, according as I find more or less facility in composing and retaining: then I call a secretary, have the windows opened, and dictate what I have composed. At the fourth or fifth hour (ten or eleven o'clock) I go for a walk in an alley, or under a portico, according to the weather, and I do not cease while walking to compose and dictate. Afterwards, I take a drive in a carriage, and here again I continue the work I busied myself with during my morning rest and my walk."[2] And he goes on giving us the account of those various days when work is mixed up with everything, even to the evening meal: for it is his custom to accompany it with instructive reading.

---

[1] Pliny, *Epist.*, I. 9. [2] Pliny, *Epist.*, IX. 36.

Even when he gives himself some unusual pleasure—for example, when he goes hunting—he is most careful to take his tablets with him. They are beside him while he sits near the nets, and, while waiting for the boar to show sport, he draws his stylet, and sets to work writing, and, if he returns empty-handed, he will at least bring back his pages full. This is not quite how we understand country life. Doubtless, not everybody was then so laborious as Pliny. There must have been people who did not always drag their secretary after them, and who, when they went hunting, left their tablets at home. But almost all were, like him, politicians, orators, men of letters, men of the world, whom fatigue had for a moment driven from town, who were preparing soon to go back to it, and who desired to profit by their sojourn in the fields, with a view to bring to their usual functions a body more robust and a mind more lively.

Knowing for whom the Roman villas were made, and what their owners went to seek in them, we find that they answered their purpose completely. Their whole construction, down to even the smallest details, was admirably thought out. Pliny the Younger has described his to us, and this account suffices to give us an idea of the others. On reading them, we shall first be struck to see what an essential resemblance these houses of Laurentum and Etruria bore to Hadrian's villa, which we have just studied. There is, in reality, only one difference between them—that which fortune and rank made between their proprietors. A mere private individual could not allow himself what an Emperor dared to do; but the general system of construction and decoration is the same, and Pliny's letters often confirm M. Daumet's restoration. I suppose that if we could see Pliny's villas, and especially that of Etruria, which was the finest, our first impression would be one of great astonishment at the multiplicity of buildings composing it. All these edifices, of different heights and forms, rather in juxtaposition than united, would appear to us more a village than a country house.[1] But we must bear in mind that a Roman had to be lodged there, and that a Roman, even when he prides himself on living simply, cannot do without a crowd of slaves. When he is not content to pack them into the cellars, but wishes, like Pliny, to give them proper chambers, which may at need be offered to his friends, there must be plenty of room and numerous separate buildings. What surprises one still more than the number of the separate buildings, is that no pains have been taken to arrange them in a regular manner; but we have already seen that the Romans, especially in their country houses, do not appear to have cared much for external appearance. Hence it is that their architects, instead of placing all the living rooms and all the chambers on the same side for the sake of symmetry, distributed them about here and there, so as to give them different aspects. They multiplied separate pavilions, in order that the occupants might be more isolated, and have a finer view from every side. The general arrangement

---

[1] I am not sure that Pliny does not mean to express a similar idea, when he says that, from his house at Laurentum, a crowd of villas are perceived " which, viewed from the sea, or even from the coast, look like a multitude of towns."—*Epist.*, II. 17.

might seem less happy, but the apartments were more comfortable, and that was enough for them. We are vain folk, who make the façade the first consideration, and provided it afford a better show we willingly submit to be badly lodged. The Romans troubled themselves less about the passers-by, and only built the house for those who were to live in it. All that could render it more pleasant to them was lavished without stint, and nothing was spared when it was a question of procuring them that fortifying rest and that variety of calm pleasures which they came to seek. Pliny was certainly far from being a voluptuary. He passed, on the contrary, for a man of the ancient manners, and the poet Sentius Augurinus sees in him several Catos.[1] One cannot, however, help feeling alarmed on seeing to what a point he pushed his search after comfort, in his pleasure-houses. One loses oneself in the enumeration which he makes of his apartments. He has dining-rooms of various sizes for all occasions. He dines in this one when he is alone; the other serves him to receive his friends in; the third is the largest, and can contain the crowd of his invited guests. The one faces the sea, and while taking one's meal one beholds the waves breaking against the walls; the other is buried in the grounds, and in it one enjoys on all sides the view of the fields and of the scenes of rustic life. Nowadays, one bed-chamber usually satisfies the most exacting; it would be difficult to say how many Pliny's villas contain. There are not only bedrooms for every want,

---

[1] Pliny, *Epist.* IV. 27 : *Ille o Plinius, ille, quot Catones!*

but for every caprice. In some one can behold the sea from all the windows; in others one hears it without seeing it. This room is in the form of an abside, and, by large openings, receives the sun at every hour of the day; the other is obscure and cool, and only lets in just so much light that one may not be in darkness. If the master desires to enliven himself, he remains in this open room, whence he can see all that passes outside; if he desires to meditate, he has a room just suited for the purpose, where he can shut himself up, and which is so arranged that no noise ever reaches his ears. Pliny calls it "his delights." In his villa he is happy to be far from Rome; in this room he seems to be far even from his villa. Let us add that these rooms are adorned with fine mosaics, are often covered with graceful pictures, and that they nearly all contain marble fountains; for water flows through it on all sides, clear, fresh, and abundant. It enlivens all by its murmur, and is one of the essential elements in the decoration of villas. It enters greatly into the capricious inventions of the architects when they wish to find new arrangements whose originality may please those fastidious and idle great lords. We remember the elegant bathing-hall surrounded by the *euripus* in Hadrian's villa. Pliny could not have such a costly edifice built for him, but he had at the end of his garden a leafy arbour supported on four columns of Carystiern marble. Under this arbour, which formed a pleasant shelter, were placed gushing fountains; a basin filled with water, constantly renewed and never flowing over; and, lastly, a couch of rest of white marble, whither one came to stretch oneself in the heat of the day. "From this bed," says Pliny, "the water escapes on all sides by little pipes, as if the very weight of him who reclines upon it caused it to spring forth."[1] To complete the whole, we must imagine baths, *piscinæ*, tennis-courts, porticoes extending in every direction for the enjoyment of all the views, alleys sanded for walks, others whose soil is firmer and more suited for riding in litters; and, lastly, for those who choose to ride on horseback, a large hippodrome, formed of a long alley, straight and sombre, shaded by plane-trees and laurels, while on all sides curved alleys wind, which cross and cut each other, so as to render the space greater and the promenade more varied. This is what was to be found in the villa of a man rich but steady, who wished, without being foolishly extravagant, to be comfortably housed in the country, in order to rest there at his ease.

We have said nothing about the parks and the gardens, which—a country house being in question—may seem strange; but it is rather difficult to speak of them. As may be well imagined, these are what have been least preserved in ancient villas. To enable us to judge of what they must have been, we have only a few paintings in which they are represented somehow or other, and what writers chance to tell us about them.

---

[1] It is a fancy of this kind which originated Varro's famous aviary, so vast, so beautiful, so full of ingenious complications, and so peopled with rare birds. The middle of the aviary formed a dining-room, where the table and couches of the guests were surrounded by running water, so that, while eating the most delicate viands, one could see the fish swim about at one's feet, and hear the blackbirds and the nightingales singing around one.

Their data are somewhat incomplete, and only half satisfy our curiosity; but they have at least the advantage of being all in agreement with each other. Among the landscapes which decorate ancient houses at Rome and at Pompeii, some pictures of gardens have been found. These are always regular alleys, shut in by two walls of hornbeam, cutting each other at right angles. In the centre a kind of round space is usually found, with a basin in which swans are swimming about. Every here and there little arbours of greenery have been arranged, formed of canes interlaced and covered with vines, at whose end a marble column or a statue is seen, with seats placed around to allow promenaders to rest for a moment.

These paintings remind one of the following saying of Quintilian, which naïvely expresses the taste of his age: "Is there anything finer than a *quincunx* so disposed that, from whatever side one looks, only straight alleys are perceived?"[1] Writers add some curious details to this information. It is seen from the descriptions of Pliny the Younger, that in his gardens, as in the paintings of which we have just spoken, the alleys were bordered by veritable walls of verdure. He thus with great pleasure describes for us a fine alley of plane-trees, of which he is very proud. "My plane-trees," he says, "are covered with ivy, which, twining around the trunks and the branches, and stretching from one tree to another, binds them all together. Between them, to make the wall thicker, box has been planted, and behind the box, laurel, in order to finish and fill up the

---

[1] Quint., VIII. 3, 9.

interval." Box, especially, plays a great part in Roman gardens. It not only forms the border of the parterre and pleasingly frames the capricious designs traced on it, but it is cut in the strangest manner. They were not satisfied with cutting it into pyramids, or forming vases of it, as has been done at Versailles; they made it represent animals looking at each other, and even fashionnd it into letters giving the name of the owner or the gardener.[1] These fancies were in vogue from the time of Augustus. It would seem as if the Romans, intoxicated as it were by their fortune, then became more sensible to what Saint Simon calls "the proud pleasure of compelling nature." At the same time that they try to introduce the country into the town, they bring the town into the fields. In order to level the ground on which they are about to erect their villas, they raise hills and fill up valleys. In their gardens they only like trees whose growth has been stunted or whose form distorted. There are, indeed, a few sensible people, and especially the poets Horace, Propertius, and Juvenal, who protest against these caprices. Seneca declares roundly that he prefers "brooks whose course has not been constrained, and which flow as nature pleases, and fields which are charming without art," but Seneca did not the less inhabit villas in the taste of the day. He had clipped hedges, cut box, counterfeited trees, and all the tricks

---

[1] Let us not too hastily laugh at this mania, for do we not see it revive before our eyes? Has it not of late become the fashion to trace strange designs in our gardens with flowers? We already form the owner's initials and shall soon get to write his entire name.

which he thought ridiculous. So true is it that it is easier to laugh at fashion than to escape from it.

Furthermore, it is evident that gardens and parks had not then the importance which they have assumed with us. This is well seen from the little room they occupy in Pliny's descriptions. The ancients did not possess all the means of varying and embellishing them known to us to-day. They lacked several of the trees which form their ornament, and, above all, their flora was not so rich.[1] Their gardens, therefore, were less capable of natural adornment than ours, and, indeed, they cared much less for it than we do. What stands them in lieu of all, and what they most passionately desire in their villas, is the view. In order to obtain a broad or smiling view, embracing a vast horizon, or looking on a charming spot, they grudge nothing. It is the chief charm of their pleasure houses. They consent to walk or ride in litters, in monotonous alleys, between two hedges of hornbeam but when they are at home, in their dining-rooms, in their chambers, or in their studios, they choose to have the most beautiful spots before their eyes. They love nature and enjoy the country from their windows, as it were.

However, we must here make yet another distinction. The views sought after by the Romans were not

---

[1] M. Friedlaender remarks that Europe owes part of the magnificent flora of its gardens to the great taste of the Turks for flowers. The tulip, the lilac, and the ranunculus, as well as the cherry-laurel and the mimosa, have passed through Vienna and Venice into the west. Later on, the discovery of America brought a new and much more abundant importation of flowers and ornamental plants into Europe.

always those preferred by us, and among the sites that please us most there are some that would not have been to their taste. Their love for nature had its preferences and its limits. Large plains, fine meadows, and fertile fields enchanted them. Lucretius can imagine no greater pleasure, on days when one has nothing to do, than "to recline by a brook of running water, beneath the leafage of a high tree," and Virgil wishes himself, as the greatest bliss, that "he may always love tilled fields and rivers that flow along the valleys." This is the foreground of the landscapes they love — meadows, crops, a few fine trees, and water. Let us add to it, as a background to the picture, some hills on the horizon, above all, if their sides be cultivated and their summits wooded. The frame is thus filled in. It only contains those simple and proportioned beauties which please these delicate artists above all things. But it must be owned that, if rich and civilised nature charms them, they have less understanding for the grandeur of nature in its wild state. Cicero says downright that only the force of habit can enable us to find a charm in mountain views. During many centuries Roman officers, chiefs of legions, governors of provinces,—people of open mind and awakened taste—crossed the Alps without feeling other sensations than tedium and fear. They would have been much surprised to learn that one day thousands of travellers would admire the spectacle which seemed to them so repellent. People went but little into the mountains from curiosity. If a journey over the St Gothard could not be dispensed with, before undertaking it, prayers were offered up to Jupiter *pro itu et reditu*, and the poet Claudianus says that when the glaciers were caught sight of, it was as if the travellers had seen the Gorgon, so terrified were they.[1] It is certainly a conquest to have become sensible to these fine spectacles, and we must congratulate ourselves on having done so, but perhaps we have lost on the one hand what we have gained on the other. I quite admit that we understand the poetry of a wild spot better than the ancients did, but, do we feel so acutely as they what St Beuve calls "the charms of a reposeful landscape"? When we go over Upper Italy, and arrive in the neighbourhood of Mantua and the banks of the Po, the view of this country, formerly celebrated among travellers, leaves us nearly indifferent. Our minds being taken up by the fine views of the Alps, which we have just crossed, we scarcely vouchsafe even a disdainful glance at that smiling meadowland and the great calm river that waters it. Yet it is the fatherland of Virgil, the country which he had before his eyes in his childhood, and which never left his heart. Those plains, to us so void of character, awoke in him the love of nature. In order to understand it, he did not need to plunge into the mountains, to climb to the region of eternal snows, and to see the great rivers issue from the glaciers. For him, "it was enough to look upon those green fields, to walk along these brooks, under the pale foliage of

---

[1] Claudienus, *De bello get.*, 340, etc. See, on this subject, the chapter of Friedlaender on the sentiment of nature among the Romans, at the end of the second volume of the French translation.

the willows, to take the shade and the freshness beside the sacred founts, and to listen at eventide to the complaint of the wood-pigeon and the distant songs of the peasant cutting his trees." It is thus that there awoke in his soul that deep feeling of universal life and that generous sympathy with nature which enraptures us in his verse. So, have we gained as much as is asserted, if, by dint of progress, we have become incapable of understanding the scenes and loving the country that inspired such beautiful works?

To return finally to the villa at Tibur and the prince who built it: it seems to me that Hadrian and his country house give us on the whole a true enough idea of the way in which Romans understood nature and enjoyed it, and that this manner is neither so unreasonable nor so different from our own as is supposed. Like the curious of our day, Hadrian roamed about the world a great deal, and visited in preference countries whose natural beauties are heightened by great historical memories. Such a taste will appear strange to none. Nature also appealed to him for her own sake, and we see that he did what was not done in his time, he climbed Mount Ætna and Mount Casius. But when he chose to construct a country house for his last years, he did not build it on the side of Casius or of Ætna; and he did well. Those are spectacles one is charmed to see once, but which it is not well to have always before one's eyes. He chose one of those more limited, less grand sites which do not crush mankind by their sublimity, which do not constantly over-excite his admiration and end by fatiguing, but which, on the contrary, induce calm and repose. In order to see whether his choice was a happy one, we need only return for a moment to the villa at Tibur, and look upon the admirable view which is enjoyed from the Pœcile. Let us place ourselves on the circular piazza that ends it, and which was so arranged that nothing of the fine spectacle might be lost. We may be sure that there were marble benches here, and that Hadrian and his friends often came hither to sit at day's decline. Before us, Rome first attracts our eyes. The whole of it is seen with its towers and domes outlined against the sky. Who knows but that Hadrian, in placing his villa opposite to his capital, wished to give himself the pleasure of a striking contrast? The poet says that there is nothing more pleasant than to hear the wind howl when one is quietly sheltered in one's house, and perhaps it seemed to this prince, wearied with power and life's activity, that this distant bustle would make his rest more sweet. But if Rome first attracts the attention, the surrounding views soon seize upon and enchain it. Near us, on all sides, the hills gradually rise and climb, growing more green and smiling as they retire farther from the plain. To the left are seen the summits of the Latian hills, to the right the picturesque mountains of the Sabina, Mentana, Monticelli, and, further off, Palombara, at the foot of Mount Gennaro. It is impossible to imagine an horizon more simple, and, at the same time, more expansive, more grand and calm, more variety and proportion. "It is not only a landscape," says Pliny the Younger, "it is a picture."[1] It is difficult to tear

---

[1] Pliny, *Epist.*, V. 6.

oneself from this spectacle, and in leaving it one tells oneself that it is impossible to maintain that people who knew so well how to choose the situation of a pleasure house did not love the country and understand nature.

CHAPTER V.

OSTIA.

To talk of Ostia is not to wander far from Rome. In spite of the distance separating them, Ostia may be looked upon as one of the outskirts of the great city. It has always been mixed up in its history, was necessary to its existence, and early became one of the organs of its life. The journey to Rome would therefore seem incomplete were one to neglect to go and see it.

Yet there is no easy way of getting there. There being no public coach running thither, it is an excursion to be thought over and prepared in advance; a circumstance that discourages many of the curious from undertaking it.[1] The first part of the journey is monotonous enough. We leave Rome by the St Paul gate, the ancient *porta Ostiensis*, and follow the Tiber nearly all the way. The banks of rivers are usually smiling and green, and we guess their course by the clumps of trees that shade it. Here there is no verdure: the Tiber, yellow and silent, flows between a few meagre shrubs and bushes whitened

---

[1] A railway has lately been constructed, which goes as far as Fiumicino; but from Fiumicino to Ostia the road is long and inconvenient. The *isola sacra*, peopled by herds of nearly wild cattle, has to be traversed, and the Tiber to be crossed.

by the dust. Yet in the flourishing times of the Empire this was a pleasure resort. Financiers and great lords bought a small garden near the Tiber at a high price. They feasted their friends of both sexes there, and a poet of the time represents them as quaffing dainty wines from goblets carved by great artists, to the jocund sound of the barques continually passing up and down the river.[1] To-day there are no longer either barques or gardens, and nothing troubles the solitude of this desert but a few droves of horses or oxen, led by hard-eyed herds, whom the passer-by scares away. We meet nothing but one or two peasants on horseback, returning from town, with their picturesque costume, their great boots, their pointed hats, and their long sticks slung across the saddle. Time slips by; the road continues to rise and fall, and the spectacle is always the same. At length, after two hours of this monotonous road, thickets are seen, trees reappear, and the horizon broadens. We behold in the distance the parasol pines of Castel-Fusano, pass a few cornfields, and soon reach Ostia.[2]

---

[1] Propertius, I. 14.

[2] The Roman administration has not yet published a map showing the present state of the excavations at Ostia. At my request, M. Laloux, a member of the Académie de France at Rome, kindly repaired to Ostia, and, taking for the basis of his work a plan of Canina, marked out the discoveries made for twenty years past. If readers, thanks to the map he has traced, follow the account which they are about to read of these discoveries with more facility and pleasure, they, like myself, will thank M. Laloux for the trouble he has taken for them.

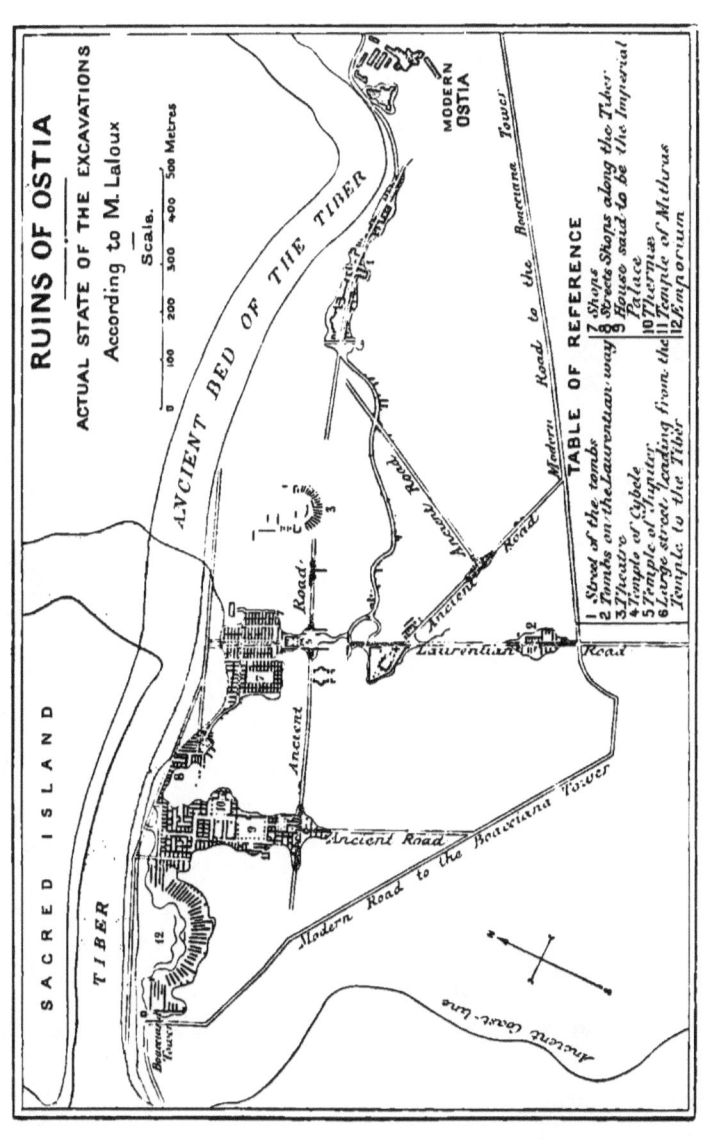

I.

MODERN OSTIA—ASPECT OF THE PLAIN BY WHICH ANCIENT OSTIA IS COVERED—HOW THE TOWN CAME TO BE ABANDONED—THE FIRST EXCAVATIONS MADE THERE—SIGNOR VISCONTI'S LABOURS—DISCOVERY OF THE STREET OF TOMBS—THE HOUSE KNOWN AS THE IMPERIAL PALACE—THE GREAT TEMPLE AND THE STREET LEADING TOWARDS THE TIBER—THE SHOPS SITUATED ALONG THE RIVER.

THE modern town presents itself to us under the aspect of a church of the sixteenth century, and of an elegant castle with the arms of Julius II. graven upon it. About the castle two or three houses are closely crowded. They constitute the entire town. During the fever season, which begins early and ends late, the inhabitants are about a dozen in number. In the month of November, some hundreds of peasants arrive from the vicinity, crowd themselves into huts, and cultivate the country. As soon as the heats return they hasten away.

On advancing a few steps beyond the houses and the château, and looking around, one is struck by the majestic spectacle before one. Not a sound rises from the great surrounding plain—all seems motionless and mute. It is a loneliness and a melancholy which strikes the soul. Our emotion is increased when we reflect that this silent spot was formerly one of the most bustling in the world, peopled as it was with the busy crowd that thronged it when the fleets of Africa and Egypt brought thither the corn which fed Rome. The sea, sparkling in the distance, forms, as it were, a luminous frame to the desolate picture. To the right the Tiber divides into two branches, which environ the *isola sacra*, now inhabited by herds of buffaloes; around us, as far as the eye can reach, the plain is covered with little hillocks of unequal size. These are heaps of rubbish covering a great buried city. Below this piled-up earth, in which our foot strikes at every step against fragments of marble, pieces of pottery, and the handles or bottoms of broken vases, we are sure to find ancient Ostia again.

Such an assertion may at first sight cause some surprise. We can understand that the eruption of Vesuvius, seizing Pompeii in the fulness of life, and in a single day burying it all beneath its ashes, should have preserved it for us as it was; but Ostia was not, like Pompeii, the victim of a sudden catastrophe; it perished slowly and in detail. How comes it, then, that it is hoped to find important remains of it again? Because it ceased to be inhabited all at once. Its prosperity depended upon the prosperity of Rome, of which it was the port, and declined rapidly when Rome no longer drew to her the travellers and the merchandise of the entire world. The invasions of the barbarians struck it the final blow. From Genseric downward, it was the natural road of all the bold pirates who were tempted by the riches accumulated in the Roman Campagna.[1]

---

[1] Even in Cicero's time, a Roman fleet commanded by a consul had been surprised and destroyed at Ostia. "Almost before the eyes of Rome," says Cicero, to whom this misfortune appears an ignominy (*Pro lege Man.*, 12). The pirates, who were kept away during the flourishing days of the Empire, returned in the fourth century.

It was there they disembarked in order to be nearer to their prey, when they wished to attempt some advantageous *coup-de-main* without giving time for preparations of defence. These repeated incursions soon made the sojourn at Ostia unbearable. The poor town was then forced bitterly to regret its neighbourhood to the sea, which, after having so long made its fortune, exposed it to so many unforeseen attacks. Every attack of which it was the victim diminished its population. Probably, one day, the last remaining inhabitants, threatened with an onslaught more furious than the others, and seized with fear, suddenly fled together far from the coast. They doubtless sought some asylum, either in the mountains of Latium or the Sabina, where they felt certain the enemy would not follow them, or behind the walls of Rome just rebuilt by the Emperor Honorius. Once out of the town, they were not tempted to return. The incursions of the pilferers became each day more frequent. It may be said that, from the last days of the Empire down to our own time, they never stopped, and that safety was never for a moment restored to this unhappy shore. The Vandals were succeeded by the Saracens and the Barbary pirates, whose unceasing depredations inspired the people of the country with a terror of which the memory has continued to live on all the maritime coast of Latium. Under Pope Leo XII., a short time before the taking of Algiers by the French, people still spoke of houses pillaged by them, and of peasants whom they had carried off in order to make slaves of them. This is why Ostia, once abandoned by its inhabitants, was never repeopled, and it is just this which has preserved its remains. Other Roman towns had, doubtless, much to suffer from the Goths, the Lombards, and the Franks; but they continued to live, and, living, renewed themselves. It being necessary to lodge, when houses became too old, they were rebuilt. The old ones furnished the materials for the new, and nothing of the ancient constructions remained. It is man, much more than time, who destroys the monuments of the past. Fortunately for Ostia, it has only had to do with time. It has doubtless often been pillaged; but the pillagers were generally in a hurry, and had not time to ravage thoroughly. Moreover, they did not care to take all. They entered the deserted houses, hurriedly loaded themselves with what seemed to them precious, and what they could easily carry away. Sometimes they violated graves, when they hoped to find rich booty there. On the road leading from Rome to Ostia, a large slab that covered one of the finest tombs was brutally prised up with a lever and flung into the middle of the road, where it has been found. Temples attracted them above all. Along the walls of the Temple of Cybele marble linings are seen in splinters, and iron clamps twisted. Below, inscriptions inform us that wealthy worshippers had here consecrated silver statues, representing emperors or gods. The inscriptions are still there, but the statues have disappeared, and this twisted iron and broken marble show us with what roughness and violence the removal was effected. But if they took the silver statues, they left the marble ones, whose value they did not know, and which would have been too cumbersome to carry off. Neither could they take away the houses. This is how, in spite of such frequent ravages, so many remains of ancient Ostia still exist. When nothing was left to tempt robbers, they returned no more, and left the town to perish from age. Little by little the walls fell in, the columns of brick and stone sank one by one, crushing each other in their fall, and then, with time, a layer of earth covered all, and grass grew over the ruins. But, below, there still exist the solid foundations of houses and public monuments, pavements of mosaic or of marble, prone columns, broken friezes, and, without doubt, fragments of wall as well, protected by the very fall of neighbouring buildings. Excavations might, therefore, be undertaken without misgiving, for, I repeat, on raising this rubbish, there was the certainty of finding the remains of a great city.

Amateurs of the last century knew this well, and therefore sounded nearly all this vast plain, each time drawing forth remarkable works of art. These fortunate discoveries, the precious marbles with which the ground is spread over, as it were, and the inscriptions everywhere found, at last aroused public attention. Many persons said to themselves that perhaps they had another Pompeii within their grasp, at but a few miles from Rome, and that this good fortune should not be neglected. In 1800 it occurred to Pope Pius VII. to begin regular excavations, which were directed by the architect J. Petrini; but, unfortunately, political events soon interrupted them. They were not resumed until 1855, when Pius IX. entrusted Signor Visconti with their prosecution. The works, carried out by convicts who had been lodged in the castle of Julius II., were well conducted, and the success obtained from the commencement drew the attention of the learned world to them.[1]

At the time when the excavations began, nothing of ancient Ostia remained upright but the four walls of a temple, called, I know not why, the Temple of Jupiter, which must have been one of the most important of the town.[2] This temple had been saved from destruction by its height, being built above a huge substruction, forming a sort of lower storey, almost as high as the temple itself. The fragments of neighbouring houses having covered up all this storey, the door of the monument became on a level with the new ground surface, and, fortune helping, the four walls had held fast. This, then, was the only building which had survived the general ruin, and it attracted attention from all sides of the immense plain. Excavations were begun on this side, in the time of Pius VII., and the vicinity of the temple cleared. Signor Visconti chose to proceed in another way, and follow a more regular course. Instead of establishing himself at once in the heart of the city which he desired to uncover, as Petrini had done, he attacked it, as it were, from the outside, and tried to enter by the gate. He remembered that at a certain spot a large number of funereal inscriptions had been found, and supposed that it must be near a public way. At Ostia, as everywhere, tombs were placed on both sides of the highways, and the abode of the living was only reached after passing through that

---

[1] Signor G. S. Visconti made known the chief results of these excavations in the *Ann. de l'Inst. de corresp. archéol.*, 1857, p. 281, etc.

[2] See No. 5 on plan.

of the dead. These suppositions were found to be correct, and on digging round tombs it was not long ere the large stones of the *Via Ostiensis* were discovered. From that moment, there was no possibility of mistake, and, in order to reach the city gate, it was only necessary to go straight on.[1]

The road was cleared for some distance. It is a way 5 mètres broad, with roomy side walks and two rows of tombs. These tombs, in general less handsome than those at Pompeii, are also of a more mixed character. Besides very simple *columbaria*, containing freedmen or poor people, is found the tomb of a somewhat vain Roman knight, who had caused himself to be represented with the insignia of his dignity and genii offering him wreaths. At Ostia a knight must have been a great personage. Then the remains of rather extensive premises are found, divided into a large number of small rooms, which, according to some, served as a barracks, and according to others, as an inn. Thence we arrive at one of the gates of the town, whose threshold is still in its place, and enter Ostia. The quarter by which we enter is wretched enough, as is usually the case with the outskirts of large towns, and especially of commercial towns, where so many poor people are crowded together. The chief street is bordered by houses of small and poor appearance, and it is soon seen to divide into several more narrow streets, leading in contrary directions Signor Visconti hesitated to penetrate further. The walls which he

---

[1] See No. 1 on plan. Other tombs, less curious, have been found along what is believed to be the *Via Laurentina.* See No. 2 on plan.

met with on his way had been repaired anyhow, with ruins brought from elsewhere, while a small stone urn, taken from a tomb, had been made into the basin of a fountain. He concluded from these signs that he had fallen upon a quarter very hurriedly reconstructed in the fifth or sixth century, after a first disaster to Ostia, when the frightened inhabitants sought to get away from the sea where so many enemies came from, and thronged into this little corner of the town, on the side towards Rome, whence aid might be hoped for.

But he had, at the same time, approached the town by another of its extremities, towards the spot where it touched the sea, and here he was more fortunate. A little lower than the *Boacciana* tower a considerable mass of ruins had long since been remarked, arranged so as to form a semicircle, and doubtless belonging to some important building. It was generally supposed that it must have been a market (*emporium*), and as Canina remembered to have seen a reproduction of such a monument on a coin of the Emperor Severus, and this prince had made a large road (*Via Severina*), which started just from the spot and skirted the whole littoral, he did not hesitate to believe that he had built this market too, and to call it *emporium Severi*.[1] Beside the *emporium*, there rose quite a hill of rubbish; Signor Visconti thought that it must cover some rich habitation, and made his workmen resolutely attack it. First a statue of Ceres was found, and then, under twenty feet of earth, the most beautiful mosaic that had been discovered in Rome for some time past. "This marble

---

[1] See No. 12 on plan.

pavement," says one of the explorers, " confirms Ennius Quirinus Visconti, who holds that in mosaics of the kind we see an imitation of the carpets of Alexandria, which were the delight of antiquity. These capricious arabesques, enclosed in regular compartments, surrounded by festoons and scroll-work of the richest invention, and set off with the brightest and most harmonious colours, produce the same effect, and have for the eye the same charm, as the most magnificent of carpets."[1] It was soon found, by unmistakable indications, that the hall where this fine mosaic was placed belonged to some baths, and, as ornaments had been lavished upon them, it was supposed that they were public ones. It was known with certainty, from a curious inscription, that the Emperor Antoninus had had sea-water *thermœ* built at Ostia, which cost him more than 2,000,000 sesterces (400,000 francs), and it was believed that they were brought to light again.[2] But on continuing the excavations, it was perceived that, notwithstanding their magnificence, these baths were only an accessory to a sumptuous dwelling which is now quite cleared. It occupies a large space, or, as the Romans used to say, an entire island, shut in between four streets. The chief entrance near the Tiber is adorned with two fine columns of *cipollino*, now replaced upon their bases. The house is built like those of Pompeii, but the peristyle is so vast, and the rooms so numerous and so large, that it is suspected that they did not serve to lodge a mere private person; and it

---

[1] This mosaic is now in the Vatican Museum.
[2] See No. 10 on plan.

being known that the Emperors often sojourned at Ostia, it has been supposed that they lodged in this beautiful dwelling, which has hence been called the Imperial Palace.[1] This hypothesis is based on no solid reason, and it seems more natural to believe that the house belonged to some rich banker or great merchant, of whom we shall presently show there was no lack at Ostia.

This quarter is not the only one where manifest traces of the importance and prosperity of the town are met with. The Temple of Jupiter, of which I have just spoken, is now entirely freed, and when it had been disencumbered of the ruins covering its base, it appeared in all its splendour. Like most of our churches of the Middle Ages, it was composed of two edifices, placed one above the other, the lower one serving as a store and magazine to the temple itself. The pediment is upheld by six Corinthian columns, of which only shapeless fragments remain. But we still possess some of the elegant sculptures that ornamented the frieze, and time has respected the threshold of the door, formed of an admirable block of African marble, 4 mètres long.[2] From this we judge of the magnificence of the remainder. Before the temple, whose entrance is turned towards the south, there extends a little piazza, which was ornamented with porticoes, and on the other side a straight street leads towards the Tiber, that is to say, towards the centre of movement and business. Like the *rue de*

---

[1] See No. 9 on plan.

[2] Among the ruins of Ostia a great quantity of precious marbles has been found. The finest have been used to adorn the confessional of St Santa Maria Maggiore.

Rivoli, it was bordered on each side with porticoes. The brick pillars that supported it have remained in their places, and it is easy in thought to people it again with the crowd of promenaders who come to shelter themselves there in the hot hours of the day. On either side we enter large magazines, of which some have been entirely cleared, and which must have been very extensive and rich.[1] This street, including the porticoes, is 15 mètres broad. It is the largest of the Roman ways yet discovered, and nothing in Pompeii gives an idea of it.

This was the condition of the works when, in 1870, a change of Government took place at Rome. The excavations at Ostia were not interrupted, but their directions merely transferred to Signor Pietro Rosa, known to the public by the discoveries he had just made on the Palatine. Signor Rosa, who is of an inventive mind and full of resources, had from the very first day a happy idea, which was bound to be fruitful of results. He did not care to continue the works of Signor Visconti, whom he replaced, but wished to try new ways and turn the excavations in another direction. He said to himself that Ostia, being one of the great commercial cities of the Empire, and receiving merchandise from all quarters of the world, must certainly possess warehouses in which to store them, and that ordinary custom and good sense both suggested that these warehouses must be situated along the Tiber. It is there he sought them, and they were easily brought to light. The Tiber here forms a semicircle, around which the town was built. All trace of the quays has dis-

---

[1] See No. 7 on plan.

appeared, and the water beats against the walls of the houses. Some even rest upon solid piles, which advance into the river, so that vessels could now enter the cellars and deposit their wares there at once. The vast vaulted magazines that received the merchandise still exist, and those large amphoræ are found there, half-buried in the earth, for the storage of grain and oil. They have been much used, and some bear traces of having been mended. All these houses open upon a street which, in the days of Ostia's prosperity, must have been very lively. It is parallel to the river, and communicates with it by lanes, or rather, small passages. One of these passages is shut by a gate of monumental appearance, a circumstance proving that even in these commercial quarters there was a certain taste for elegance, and that business cares were mingled with a feeling for art. The street of the docks, as it might be called, has been freed throughout a great portion of its length, and can now be followed as far as the market of Severus.

II.

WHY THE PORT OF OSTIA WAS FOUNDED—THE FREE DISTRIBUTION OF CORN IN ROME—THE DIFFICULTY OF PROVISIONING ROME—CREATION OF THE PORT OF CLAUDIUS—THE PORT OF TRAJAN—THE IMPERIAL PALACE—THE TOWN OF PORTUS—THE MAGNIFICENCE OF OSTIA AND PORTUS.

WHILE traversing this long street, and passing between these two rows of storehouses, broken from time to time by glimpses of the Tiber, we find ourselves transported into a world of industry and trade, which shows us antiquity in a new light. Ancient historians scarcely speak to us of the economic circumstances of the communities of their times, and seem not to suspect that one might some day be curious to know how those communities obtained their subsistence; how they exchanged their merchandise for that of their neighbours, and whence came the objects necessary or agreeable to their lives. These details appear to them too low, and liking to show us their epoch only by its nobler sides, they do not care to descend to them. It is at Ostia, especially, that such questions suggest themselves, and it is also there that they are most easily solved. The sight of its ruins and the memories of its history may give us more than one useful hint.

Tradition attributes the founding of Ostia to a Roman king, Ancus Martius. "It is he," says the old poet Ennius, "who built this harbour for the beautiful ships, and for the sailors who risk their lives upon the waves."[1] When Rome had become mistress of the world, the sages who sought to discover the reasons that had made her so powerful, congratulated Romulus on not having placed his town on the sea-shore. Cicero, after the Greek philosophers, enumerates all the dangers to which maritime towns are exposed. He tells us that nothing can warn them of the surprises of an enemy, who may land on the coast and penetrate within their walls without anybody having suspected

---

[1] *Ostia munita est : idem loca navibus pulchris Munda facit ; nautisque mari quærentibus vitam.*
—*Ennii relig.*, Vahlen, p. 24.

his approach. He adds that they are more accessible to outside influences, and without defence from the corruption of foreign manners. Those who inhabit them do not attach themselves to their homes. A continual change of desire and hopes carries them far away from their country, and even when they do not change their abiding place, their always adventurous mind travels and runs about the world.[1] It is this that lost Carthage and the beautiful Isle of Greece, which, in the midst of this belt of waves, seem also to swim with the institutions and manners of their unstable cities. Hence, Cicero infers that Romulus gave proof of rare sagacity by establishing himself in the interior of the land, and yet near a river which could bring him the wares of neighbouring countries. Whether the founder of Rome went through all the fine reasoning attributed to him is very doubtful; but it is certain that it was beneficial for the new town not to be too far from the sea, and that she soon sought to profit by its advantageous vicinity for the advancement of her fortune. Her citizens were animated by passions which at first sight appear incompatible. They are usually shown under only one of their aspects—the finest and most brilliant. They were soldiers and conquerors, to whom tradition now only attributes heroic attitudes; but among these demigods there were traders and usurers. They were as greedy as they were brave. They loved glory; but they were also very partial to money. They knew very well how to calculate and, under a disdainful exterior, took great care not to neglect the good profits to be drawn from com-

---

[1] Cic., *De Rep.*, II. 3.

merce. It was in order to please them that Ancus Martius founded the port of Ostia, where the Tiber throws itself into the sea.

A king of Rome, at that period, was not rich enough to undertake costly works at a distance. The foundation of an arsenal (naval) is attributed to him, but he probably built neither harbours nor jetties. At least no traces of any have been found.[1] The river's mouth itself formed the port, and not much pains was taken to render it more commodious and safe. Such as it was, it served throughout the Republic. In its small and shallow space, it sheltered not only the mercantile but the state navy. Titus Livius informs us that, during the Punic Wars, several squadrons left Ostia to attack the fleets of Carthage. Yet it was not possible always to be content with the old harbour of Ancus Martius. Not only must it necessarily become insufficient as Rome's commerce grew with her power, but the Tiber ere long began to silt up its approaches. The "yellow river," as it was called, carries large quantities of mud along with it, and Signor Lanciani calculates that, at the Fiumicino mouth, the shore advances into the sea more than 3 mètres every year, and at the Ostia mouth 9 mètres. Access to the port therefore became more difficult every day, and, towards the end of the Republic, large ships could scarcely enter it any longer.

But this was, nevertheless, just the time when Rome

---

[1] Some remnants of tufa and travertine construction near the house called the Imperial Palace, on the side of the Boacciana tower, have been assigned to the *Navalia* of Ostia. But these remains, whatever monument they belong to, must be of the last century of the Republic. See *Ann. de l'Inst. de corresp. archéologique*, 1868, p. 148.

for her subsistence most needed to draw to her the vessels of the entire world. How was it that the Roman Campagna, that rich and well-cultivated ground, so soon became unable to feed its inhabitants? Pliny the Elder especially accuses the increase of large proprietorship: *Latifundia perdidere Italiam.* These wide domains, which had absorbed the heritage of so many poor families, contained parks, gardens, porticoes, and promenades; all so much withdrawn from agriculture. In the remainder of the land the owners were everywhere induced to replace corn-fields by pastures, which produced a more certain revenue, and are more easy to work. M. Mommsen adds that foreign competition discouraged Roman agriculturists, and that when they saw the merchants of Sicily and Egypt bring the wheat of their country in abundance and at a low price, they ceased to grow it. From that moment, Rome, powerful Rome, was at the mercy of her neighbours, henceforth living only on produce from without, which the sea brought her through a thousand dangers. "Every day," says Tacitus, in his energetic language, "the Roman people is the sport of the waves and the tempests."[1] At the same time, and as if to make the evil incurable, the chiefs of the democracy, at length risen to power, paid the people for their kindness by a liberality whose consequences were necessarily fatal to the Republic. G. Gracchus caused it to be decided that henceforth the State should undertake to partially feed the poor citizens. Corn tickets

---

[1] *Ann.*, III. 54: *Vita populi romani per incerta maris et tempestatum quotidie volvitur.*

were distributed to them (*tesseræ frumentariæ*), which allowed them to receive it at half price. It being natural not to stop at half measures, some time after the Gracchi it occurred to another demagogue to give it for nothing. The less people paid, the more the number of those who desired to enjoy this favour increased. When Cæsar took possession of the supreme power their number mounted to 320,000. Popular though he wished to be, he found this a great deal too many, and reduced it to 150,000, still a fairly large number. It is said that Augustus would have gone further, and for a moment thought of giving nothing to anybody. Suetonius reports that, after a famine in which all troops of vendable slaves, all bands of gladiators, and all foreigners, with the exception of professors and physicians, were driven out of Rome, the Emperor thought of suppressing the gratuitous distributions entirely. He saw that they encouraged idleness, and caused the fields to be deserted. He retained them, however, for, says his historian, he feared lest, if he discontinued them, some ambitious person might acquire the favour of the people by promising their re-introduction.[1] He even ended by being less strict than Cæsar, and at his death 200,000 citizens were receiving corn from the State.[2] This is much, if we reflect that in Paris only 113,000 persons are inscribed on the list of

---

[1] Suetonius, *Aug.*, 42.

[2] This number was continued until the Severus dynasty. On all questions concerning the distribution of corn, the reader may consult a very complete work of M. Otto Hirschfeld, entitled *Die Getraidcverwaltung in der römischen Kaiserzeit*, which appeared in the *Philologus*, in 1870.

the *Assistance Publique*; that, according to the most favourable calculations, the population of Rome was less by a good third than that of Paris, and that this population was composed in great part of slaves who had to be fed at their master's expense. Hence we should be forced to the conclusion that there were a very considerable number of poor in Rome, were it not more natural to think that many of those who came to receive the prince's alms were not veritable poor, but small citizens who were very glad to get this additional revenue to enable them to live more at their ease. They were not at all ashamed of it. On the contrary, they appear to have been proud, since these liberalities were only granted to people who enjoyed the right of citizenship, and we see some who even put in their epitaphs " that they shared the distribution of corn," in order to establish that they are citizens.

Henceforth, the provisioning of their capital became the Emperors' great care. The Roman people, so submissive, so complaisant, and so ready to flatter all the caprices of its masters, now only showed energy when it feared to see its ration of wheat curtailed. At the least delay in the monthly distributions, this populace, which usually took everything without complaining, mutinied before the palace, or, if the prince was absent, proceeded to pillage the house and break the furniture of the prefect of Rome. When it was rumoured that bread might fall short, one of those mad panics ran through the town, such as were seen in the worst days of our Revolution, and which inclined the crowd to every excess. The Emperors neglected nothing in order to forestall these fears, and, by all kinds of privileges, encouraged the merchants of every nation to bring their grain to Italy. Claudius insured great advantages to those who built vessels with this object, adding to their gains and promising to make good their losses.[1] All in any way employed in the provisioning department of Rome (*Annona*) were exempted from every other office. "They worked," said the law, "in the public interest."[2] This department was the object of so many distinctions and favours on the part of the Government, that it got to be much respected in the provinces; the feeling of its importance was general, and as its object was to enable the "sacred city" to live, it was sometimes called *Annona sancta*.[3] Cereals reached Italy from all the countries of the world; but Egypt furnished the greater part, more than half of what was consumed in Rome. This enormous quantity of corn, collected in foreign countries by the agents of the *Annona*, was sent to Italy in a special fleet, at what was judged the most opportune moment. But, as in Egypt the harvest depends upon the overflowing of the Nile, and is not always equally abundant, it occurred to Commodus to insure against this unfortunate chance by creating a new fleet, which every year went off to Carthage in search of the corn of Africa.[4] Thus the two most fertile countries of the earth were laid under contribution. Still, this was not yet enough. Egypt and Africa might both be struck with sterility at the same time, and precautions had to be taken against a

---

[1] Suet., *Claud.*, 18. [2] Dig., *L.*, 6, 5, 3.
[3] Orelli, 1810. [4] *Hist. Aug. Comm.*, 17.

general scarcity, and insure Rome against a famine which would have affected the whole world. To this end, immense granaries were built, which in time of plenty were filled against bad years. The prudent princes were careful to keep them always full, shutting up in them, we are told, enough to enable the whole populace of Rome to live during seven years. No less was needed, in order to reassure this crowd, so easily terrified, and so fearful of starvation.[1]

What explains their alarm is the circumstance that the greater part of the corn which provisioned Rome could only come to it by sea, for the sea dismayed the Romans. These valiant soldiers were not equally intrepid navigators; like the Greeks they were inclined to exaggerate the dangers of the perfidious element, and always trembled for the fate of those precious vessels which bore their means of life, and which had to cross the sea. So the appearance of the Egyptian fleet in sight of the coasts of Italy was every year an event. Seneca related that when, at Puteoli, those light craft called "the messengers" were seen, which announced the others, Campania rejoiced. The crowd thronged upon the jetties of the harbour, and sought to distinguish amid the waves of the sea, and among the multitude of the ships those of Alexandria, known by the peculiar form of their sails.[2] It was much to have crossed the Mediterranean and reached Puteoli; but the voyage, however, was not over. They had to proceed from Puteoli to Ostia, skirting the shores,

---

[1] *Hist. Aug. Scpt.*, Sev. 8; Heliog., 27.
[2] Seneca, *Epist.*, 77.

which was attended with much danger, and even when they were opposite the Tiber, and in view of Ostia, all was not past. The entrance to the river was so difficult, the shore so bad and so changing, that more than one vessel was miserably wrecked there. Had not two hundred vessels one day been seen to perish at the same time, in the port itself, where they were not protected from the tempest?[1]

This latter peril might at least have been prevented. It would have been sufficient to build at Ostia a safer harbour, easy of access for ships, and out of the reach of storms. Cæsar, it is said, thought of making it, but death prevented him, and the project was abandoned for more than a century. It was Claudius, the foolish Claudius, who had the honour of carrying it out. This poor prince, whom his domestic misfortunes have rendered ridiculous, and whose head was not very sound, was nevertheless endowed with the taste for useful works. His zeal was in this case stimulated by a personal danger undergone by him at the beginning of his reign. When he attained the Empire, Rome was cruelly suffering from a famine of which his predecessors were accused of being the cause. Caligula, who, for his part, was quite mad, had taken it into his head to ride on horseback in the gulf of Naples. In order to satisfy him, all the vessels and ships that were in the ports of Italy had been hurriedly assembled; then, by joining them together, a large bridge had been formed from Puteoli to Bauli, with taverns on the way to stop at, and the

---

[1] Tacitus, *Ann.*, XV. 18.

Emperor had enjoyed his caprice. But the vessels employed in Cæsar's pleasures had not been able to go and fetch the corn of Egypt and Africa at the favourable time, and Rome lacked bread.[1] Caligula dying in the interval, the people in their anger blamed Claudius, who was not in fault, and were near making him pay for the follies of his predecessor. Assailed in the midst of the Forum, insulted and beaten, he only escaped from the hands of these madmen, thanks to a private door which happened to be open, and enabled him to return to the Palatine.[2] That day Claudius was very frightened. So, in order to make the arrival of the grain easier, and be no more exposed to seditions of this kind, he resolved to build a new harbour at Ostia. It is said that the engineers, contrary to their custom, exaggerated the expenses of the undertaking, in order to dissuade him from it.[3] But, against his wont, he held out against everybody, and, lest the works should be negligently conducted, decided to overlook them himself. During all the time they lasted he made many stays at Ostia. He was there the day when his wife, Messalina, took it into her head to be married with great ceremony to her lover, Silius, during the life and reign of her husband. Tacitus reports that, on the morrow of the wedding, while she and her friends were

---

[1] Suet., *Calig.*, 19 ; Aurel. Vict., *Claud.*

[2] Suet., *Claud.*, 18.

[3] There must have been an important debate on this subject in the Senate. Traces of it are found in *Quintilian*, III. 21, and II. 8. All relating to the ports of Claudius and Trajan has been studied with great care by Signor Lanciani in an important article, *sulla citta di Porto* (*Ann. de l'Inst. de corresp. arch.*, 1868).

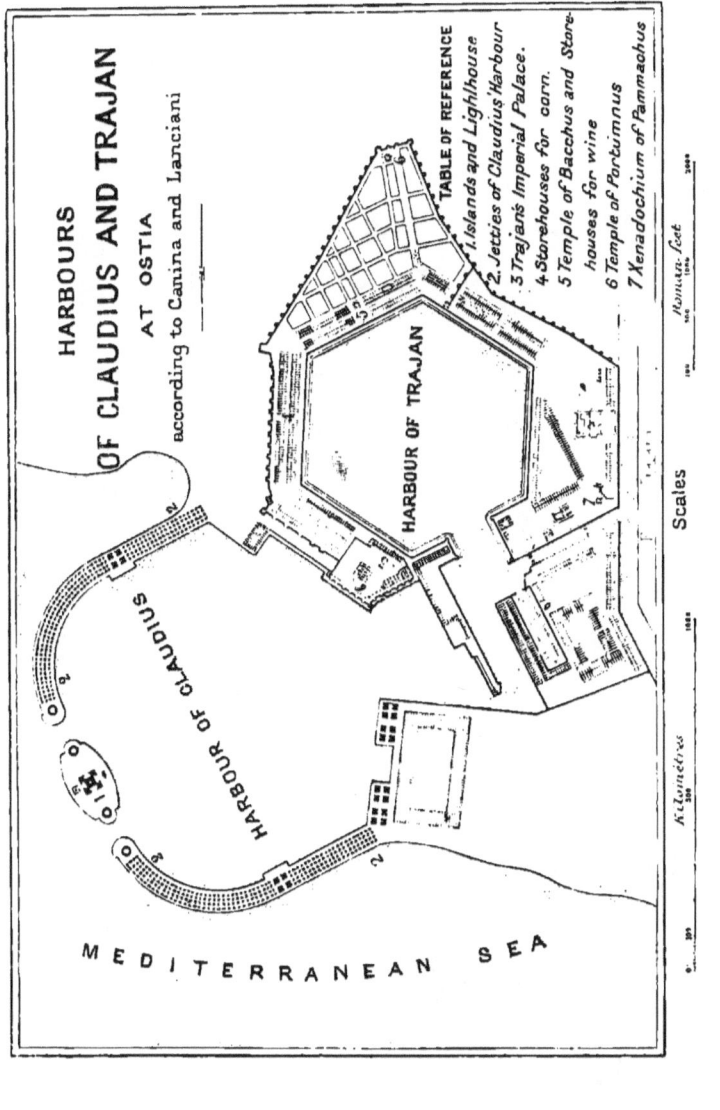

engaged in a kind of crazy, or furious bacchanal, one of them, for a drunken freak, climbed up a high tree, and when the others asked him what he saw, replied that there was a dreadful storm coming from Ostia.[1] It was the husband, who, warned a little late, came to trouble the feast.

The port of Claudius still exists,[2] only, owing to the progress of the sand, it is now quite inland. Its form can, however, be distinguished, and its extent measured. It was dug at some distance from Ostia, and above the mouth of the Tiber, perhaps with the idea of preventing it from being silted up. It was shut in to the right and to the left by two solid jetties, "like two arms," says Juvenal, "stretched out in the middle of the waves."[3] The one to the right, sheltered by its position from tempests, was formed of arches, which allowed the water of the sea to enter, while that to the left was of solid, stout masonry. It had to be strong enough to resist the billows, when raised by the south wind. Between the ends of the two jetties the enormous vessel on which one of the largest obelisks of Egypt had just been brought over, was sunk full of stone. It became a kind of islet, protecting the harbour, and only leaving on either side two narrow passages, furnished with iron chains.[4] On this little

---

[1] Tacitus, *Ann.*, XI. 31.
[2] For the plan of the ports of Ostia the map published by Canina has been used (*Atti della Pont acc. di arch.*, VIII.). But as regards the port of Trajan, care has been taken to correct Canina's plan by that of Signor Lanciani, which is more exact (*Monum. del 1st*, VIII. 9.
[3] Juvenal, XII. 77. See No. 2 on plan.
[4] See No. 1 on plan.

island a lighthouse was raised—that is to say, a tower of several stories, ornamented with columns and pilasters, like the one that lit the port of Alexandria. By the light of the rays thrown by the beacon upon the waters, the ships would direct themselves at night, and enter the port at all hours and in all weathers.

Although, according to M. Texier, the harbour of Claudius measured 70 superficial hectares, it soon became too small, and, under Trajan, it was found necessary to enlarge it. That indefatigable prince, who filled the world with buildings of all kinds, and especially useful ones, had given great thought to maritime constructions. He had repaired the harbour of Acona, and founded that of *Centumcellæ* (*Civita Vecchia*). At Ostia, instead of being content to extend the harbour of Claudius, he had a new one dug, which, like the other, is still visible inland, and whose form and outlines continue to be visible by the undulations of the ground. It was a hexagonal basin, nearly 40 hectares in extent, lined on all sides by a quay 12 mètres broad, with granite posts to moor the ships to. These are still in their places. The new port formed a continuation of the old one, to which it was joined by a canal 118 mètres broad. In order to put them in communication with the Tiber, and, by the Tiber, with Rome, another canal (*fossa Trajana*) was dug, which has in course of time become a new arm of the river, the only one now navigable, and known to us as the *Fiumicino*. Ships, therefore, entered the harbour of Claudius, and thence passed into that of Trajan, which formed a sort of inner basin. There, if too large to navigate the Tiber, their merchandise was unloaded and transported on smaller craft. A curious painting, found at Ostia itself, in the tomb of a rich shipowner, shows us how the operation was effected. This painting represents one of those vessels used for the navigation of the Tiber, and called *naves caudicariæ*. Each of them, like those of to-day, had its name by which it was called, and which was inscribed in red or in black on some conspicuous part of it. This one had received the name of a divinity, to which, in order to avoid confusion, had been added that of the proprietor. It was called the Isis of Geminius (*Isis Geminiana*). On the poop, above a little cabin, the pilot Pharnaces grasps the helm. Towards the middle, the captain Abascantus is overlooking the workmen. On the shore, porters, bending beneath the weight of sacks of corn, proceed towards a small plank, which joins the ship to the shore. One of them has already arrived, and is pouring the contents of his sack into a large measure (*modius*), while in front of him the *mensor frumentarius*, charged with the interests of the department, is intent on seeing the measure well filled, and holds the sack by its edges, in order that nothing may be lost. A little further, another porter, whose sack is empty, is sitting down to rest, and his whole face breathes an air of satisfaction, explained by the word written by the painter above his head, "I have finished" (*feci*). It is a scene of striking variety, like those every day witnessed in our seaports. The vessel, thus laden, proceeded by the *fossa Trajana* towards the Tiber, and followed the river to Rome.

With the new ports a new town came into existence. It was called, from the name of its founders, *Portus Trajani,* or simply *Portus* (now *Porto*). It must have been inhabited chiefly by merchants and officials of the *Annona*. Signor Lanciani affirms that more than two-thirds of the houses of which any remains are left were warehouses. They stretch in several rows around the basin in long regular lines, and appear to have been constructed at the same time and upon the same model. They must have had two floors, the lower one, where corn, wine, and oil were stored, and the upper one, now destroyed, doubtless containing the lodgings of the workmen and the officials. The corn stores are still recognised by the thickness of their walls and by the care taken to line them with a strong plastering, to preserve them from the damp, which was so much to be feared in such a marshy country.[1] It is thought that the warehouses of the wine merchants were situated near the Temple of Bacchus, remains of which have been recovered.[2] There must have been others for oil and marble, since there was a great commerce in them at Ostia. Besides these large depôts, indispensable to a seaport, Trajan did not omit to erect edifices destined for the embellishment of the town—baths, porticoes, and temples.[3] Lastly, being proud of his work, and having often to visit it, he built for himself a magnificent palace in grounds separating his harbour from that of Claudius.[4] Had this palace been cleared in an intelligent manner, and care been taken to preserve its

---

[1] See No. 4 on plan. All these details are drawn from the work of Signor Lanciani, cited above.

[2] See No. 5 on plan.

[3] At the entrance of the town the ruins of a temple of Portumnus are thought to have been recognised. See No. 6 on plan.

[4] See No. 3 on plan.

remains, it would doubtless have been one of the most
curious remains of Roman antiquity. M. Texier, in an
interesting article, relates how he came upon it, where its
existence was almost unknown.[1] A workman in pursuit
of a badger, seeing the animal enter a hole, pushed in a
stick in order to reach it. He soon perceived that the
opening could be easily enlarged, and when he had removed a few great stones, he saw that it gave access to a
spacious hall. M. Texier, on being apprised of what had
occurred, was the first to enter, and witnessed a fine spectacle. Within this first ray of light, penetrating depths
where darkness had reigned for centuries, fluttered a
whole world of insects who had taken up their abode
there; it illumined the bindweeds and the stalactites
hanging from the roof, and the little pools of water
shining on the floor. This hall led to another, and that
to others again. There were so many, says M. Texier,
and they were so vast, that, in order to find one's way
in the darkness, it was necessary to guide oneself
by means of a compass, as one does in a virgin forest.
From this time forth excavations were executed in the
palace of Trajan, by order of Prince Torlonia, to whom
all the land belongs; but, unfortunately, not in a scientific
interest. As only objects of art were sought after, to enrich the museum of the Lungara, the excavating was done
in great hurry and secrecy. The harvest once reaped,
they hastened, according to ancient custom, to cover up
again what had been brought to light. Signor Lanciani,

---

[1] This article was published in the *Revue generale d'architecture de Daly*, XV. M. Texier was charged by the French Government to study the alluvium of the great rivers of the Mediterranean.

who was allowed as a great favour to inspect these fine ruins, had not even leisure to design their plan. He tells us of baths, of temples, of splendid halls, and of a small theatre, completely visible, where Trajan doubtless came to refresh himself with the sight of the pantomime which he was accused of being too fond of; and, lastly, of an immense portico, whose columns, still in their places, caused the whole palace to be called in the country round, the *Palazzo delle cento Colonne*. These remains were so fine that they drew cries of admiration from the rough peasant who guided Signor Lanciani. After having escaped the barbarians of the Middle Ages, and the amateurs of the Renaissance—often more terrible than the barbarians—they ended by perishing obscurely in our own days, by the order of a great lord, clumsily enamoured of antiquities. *Quod non fecerunt barbari fecerunt Barberini.*

It was not only the emperor's palace which displayed such magnificence; we know that the two cities, Ostia and Portus, were rich and sumptuous. This is sufficiently shown by the fine columns and the admirable statues found there. Tacitus relates that, after the burning of Rome, under Nero, temporary shelters were hurriedly run up in the Field of Mars and the public gardens for the crowd of people who had no longer an asylum. These had to be provided, as quickly as possible, with furniture, and it was sent for from Ostia.[1] The town, therefore, possessed much more than was needed for the use of the inhabitants. After Nero's death, its prosperity increased still further. Independently of

---

[1] Tacitus, *Ann.*, XV. 39.

the great works of Trajan, of which I have spoken, Hadrian and Antoninus embellished Ostia with magnificent monuments; Aurelian had a new Forum built there, and the weak Emperor Tacitus gave it, out of his private fortune, one hundred columns of Numidian marble, twenty-three feet in height[1]—a very extraordinary piece of liberality at so unfortunate a time. As happens in all great industrial towns, corporations were very numerous in Ostia. The commercial world was divided into trade associations, having their places of meeting, their treasury, and their magistrates, and among these societies there were some which appear to have been very important. Naturally, very large fortunes were made, and some of those fortunate merchants, who were enriched by their dealings in corn or oil, chose to leave great mementoes of themselves. After winning opulence, they desired to obtain consideration, and displayed fabulous generosity in the embellishment of their city or the amusement of their fellow-burghers. Of such was that Lucilius Gamala, who probably lived under the Antonines, and of whose liberalities certain inscriptions inform us.[2] He was of an ancient family, and his ancestors during several centuries filled the most honourable functions at Ostia. So they had made him a decurion or municipal councillor from his cradle. Later, he became pontiff,

---

[1] *Hist. Aug. Aurel.* 45, Tac. 10.

[2] Such is, at least, the opinion of M. Mommsen, the last authority by whom the two large inscriptions concerning Gamala have been studied (*Ephemeris epigr.*, III. p. 319). MM. Visconti, Wilmanns, and Homolle had raised doubts as to one of them; M. Mommsen thinks them both authentic.

quæstor, edile, duumvir—in fine, everything it was possible to be in a Roman colony. After his death, he was decreed a public funeral, and statues were raised to him; but, on the other hand, by how many benefits had he not paid in advance for the honours with which they loaded him? The list, doubtless incomplete, is truly incredible; he had given public games, gladiatorial combats, finer and more costly than was usual, without choosing to accept the sum of money allowed by the city to the magistrate in order to help him out with his expenses. He had twice invited all the inhabitants of Ostia to dinner, and on one occasion he had even feasted them in 117 dining-rooms.[1] He had at his own expense paved a street near the Forum, in the space extending between two triumphal arches; he had repaired the temples of Vulcan, of the Tiber, and of the Castors, and had rebuilt those of Venus, of Fortune, of Ceres, and of Hope; he had presented public weights to the market and the wine exchange, and raised a marble tribunal in the Forum, and he had built an entire arsenal and restored the baths of Antoninus, which had been destroyed by fire. Finally, when in a moment of distress the city, which had undertaken to furnish the State Treasury with a considerable sum, found it difficult to keep its engagements, and was forced to sell communal property, Gamala came to its assistance, and gave it, in a single donation, 3,000,000 sesterces (600,000 francs). What an immense fortune do these liberalities

---

[1] Plutarch informs us that Cæsar, after his triumph, dined the people of Rome in 1022 dining-rooms. We thus see that Gamala imitated great examples.

imply? Such are the personages who lived in the fine houses which we discover at Ostia. It is not difficult to understand that they should have made them so magnificent and filled them with such fine works.

III.

THE RELIGIOUS MONUMENTS AT OSTIA—INTRODUCTION AND SWIFT PROGRESS OF CHRISTIANITY — THE "*XENODOCHIUM* OF PAMMACHIUS"— PRELUDE OF THE *OCTAVIUS* OF MINUTIUS FELIX — DEATH OF ST MONICA.

A CIRCUMSTANCE which strikes all who busy themselves with the antiquities of Ostia is the great number of temples and sanctuaries of every sort that were built there. Historians and inscriptions mention a great number of them, and some have been found again in excavations lately made. Ostia was evidently a devout city. It had a local *cultus*, to which it appears to have been much attached—that of Vulcan. In Ostia the pontiffs of Vulcan are chiefs in religious matters. They supervise other worships, and grant private persons who desire it leave to raise monuments in the sacred edifices. But Vulcan is not the only god worshipped there. The others are also devoutly prayed to, and especially Fortune and Hope, genuine traders' divinities; Castor and Pollux, protectors of seamen; and Ceres, who must have counted many adorers in a city enriched by traffic in corn. Foreigners, who formed a great part of the population, naturally brought their deities with them, and these enjoyed great credit. Intercourse with Egypt being very frequent, altars and statues were raised to Isis and Serapis. The Asiatic worship of the Mother of the Gods was also in great esteem, and the inhabitants of Ostia once enjoyed the sight of one of those solemn sacrifices, called *Taurobolia*, in which an important personage of the town, placed in a sort of cellar whose ceiling was pierced with a number of holes, had himself sprinkled with the blood of a bull immolated above him, which was to purify him of his sins, and assure the welfare of his family and of his city. We have still the inscription destined to preserve the memory of this religious festival. One of the most curious discoveries resulting from the excavations of recent years is that of the temple of the Mother of the Gods, beside which was found the hall of meeting of the religious corporation of the *Dendrophori*.[1] Mithra, the invincible sun, the unseizable god (*deus indeprehensibilis*), as his adorers at Ostia called him, was also the object of great homage. It is known that this worship, which excited piety by its secret associations and its mysterious sacrifices, in the last days of the Empire attained to great importance, and that all the vitality of paganism seems then to have been gathered up in it, in order to combat the new religion. Not only have numerous remains of Mithraic monuments been discovered at Ostia, but a temple consecrated to the Persian divinity. It is a species of domestic chapel in the fine house of which I spoke further back.[2] It is divided into three parts—

---

[1] See No. 4 on plan of the excavations of Ostia. This temple to the Mother of the Gods, or Cybele, was the subject of a long work by Signor C. L. Visconti in the *Annales de corresp. arch.*, 1868, p. 362.

[2] See No. 11 on plans of the ruins of Ostia.

not by columns, as in the case of Christian basilicas, but by differences of level. Each of these was doubtless reserved for the faithful of a different rank, and such a kind of classification was natural in a religion where the hierarchy had so much importance. The chapel must have been very elegant, if we may judge from the precious marbles with which it was paved. Facing the entrance door was the altar, raised four steps above the ground, with the genii who represent the two equinoxes, one holding a torch upright, and the other a torch reversed. Above the altar was placed, according to custom, the image of a young god with the Phrygian cap on his head sacrificing the bull. Some fragments of it were found on the ground. An inscription informs us that the decoration of the altar was executed at the cost of C. Cælius Hermeros, priest of this sanctuary.

Ostia therefore appears to have been a soil quite prepared in advance for Christianity. It is known that the most religious countries are those where it most quickly established itself. Seaports, transit and trading towns, where people of all countries met, where temples rose to all the gods, and where the religion of the East counted most believers, were particularly favourable to it, so its progress was probably very rapid at Ostia.[1] It soon possessed two episcopal sees, one at

---

[1] The Jews, who slipped in everywhere, must have been very numerous at Ostia and Portus. Several Greek inscriptions have been found, bearing the seven-branched candlestick and the formula "'Εν εἰρήνη." One of them mentions the chief of a community, to whom it gives the name of "Father of the Hebrews." The presence of the Jews at Ostia also explains why Christianity spread there so quickly.

Ostia itself and the other at *Portus Trajani*, which was rendered illustrious by St Hippolytus. About the time of Theodosius, a friend of St Jerome, the rich and noble Pammachius, conceived the generous idea of building at Portus a refuge for poor travellers. People who had come from Rome and were awaiting a favourable wind were received, as were those proceeding from all parts to the great city to transact their business and seek their fortune. They were so happy to find an asylum where they could rest a few days after the fatigues of their voyage, that the fame of the refuge of Pammachius soon spread throughout the entire world. St Jerome says that Britain heard of it, and that it was spoken about by the Egyptian and the Parthian.[1] Signor Rossi thinks he has found it again among the ruins of Portus.[2] Considerable remnants remain, among which a basilica and a large court, surrounded by columns taken from other monuments, are distinctly recognisable. In the fourth and fifth centuries this was the usual manner of building, and the only way of making new edifices yet known was by despoiling the old ones. As happens in the cloisters of the Middle Ages, in the centre of the court there was a kind of cistern or well, which bore an inscription, now much mutilated, but in which these words may be read: "Let him who thirsteth come hither and slake his thirst."[3] For us the Christianity of Ostia remains associated with two important

---

[1] St Jerome, *Epist.*, 77. 10.
[2] See No. 7 on plan of imperial harbours.
[3] Rossi, *Bull. di arch. crist.*, 1866.

memories which, when visiting these ruins, it is impossible to forget: the Prelude of the *Octavius* and the death of St Monica. The *Octavius* is the first attempt at a Christian apology, written by a Roman in the language of Rome, and is to-day still one of the most interesting works that can possibly be read. Its author, Minutius Felix, was a lawyer and a man of the world, who doubtless lived, and must have enjoyed himself, in an elegant society. He addresses the lettered and the worldly, and desires to make himself heard; so he takes care not to present his opinions in an arid, dogmatic form that might repel the indifferent. He gives them an agreeable turn, and seeks to pique the curiosity of his readers by a dramatic setting. His book is a dialogue, in which he pits against each other, not theological controversialists, but honest folk amusing themselves on a day of leisure. He supposes that one of his old friends Octavius, a Christian like himself, comes to see him after a long absence, and that, in order to be freer, and belong to each other more, they leave Rome for a few days, in company with a mutual friend, Cæcilius, who has remained a pagan. This is during the vintage holidays, a time when, the tribunals being closed, lawyers are in vacation. So all three start for Ostia, "a charming spot," where the mind enjoys repose and the body finds health again. One morning, as they are wending their way towards the sea, "yielding themselves up to the pleasure of treading the sand, which gives beneath their feet, and of breathing that light breeze which restores vigour to the tired limbs," Cæcilius, the pagan, seeing a statue of Serapis, salutes it, according to custom, by raising his hand to his lips. This religious act wounds Octavius, who cannot forbear saying to the other Christian: "It is ill, my brother, to leave a faithful friend in this gross error. Will you allow him to throw kisses to statues of stone, which do not deserve this honour, all covered though they be with garlands and sprinkled with oil?" At first no one answers, and the walk is continued. When one has visited the beach at Ostia, it is easy to re-tread in thought the way the friends must have pursued together. They doubtless traversed that long street skirting the Tiber, or some other parallel to it. Then, having reached the spot where the houses left off and nothing restricted the view, they enjoyed the sight of the immense horizon. They walked on the wet sand along the shore, among the boats that had been drawn up upon the beach, beside children amusing themselves by making pebbles rebound upon the water. The two Christians, whose souls are tranquil, give themselves up entirely to the pleasure of these sights; but Cæcilius looks at nothing; he is silent, sombre, and preoccupied; the few words he has just heard trouble him; he wishes for an explanation; he asks to be enlightened. Then all three sat down on the great stones that protect the jetty, and, facing this calm sea, under this bright sun, they begin to commune together of those great questions which were troubling the world. Is Minutius indeed telling us a romance? In any case it is a romance much resembling truth. I doubt not but that many a conquest made by Christianity in the second century was brought about by similar incidents, and that often a word, flung as it were by chance, touched a well-inclined soil, which finally surrendered after a few conversations such as those that took place on the beach at Ostia, and have been reported for us by Minutius.

The death of St Monica is the other great memory recalled by the ruins of Ostia. St Augustine has related the circumstances of it in one of the finest passages of his *Confessions*. Brought back, after terrible struggles, to the faith of his mother and of his youth, he had just received baptism at the hands of St Ambrose. Being resolved to break altogether with the world, and wishing to leave for ever that chair of rhetoric he had been at first so proud of, he had warned the Milanese "to seek for their children another vendor of words." He was returning to Africa with his mother, and waiting at Ostia for favourable weather for the passage. Augustine, who was poor, had probably taken lodgings in some inferior hostelry in the middle of the old town. Maybe it was only the rich who could have their dwellings built in the favoured sites skirting the shore. He only speaks to us of a window looking out upon a peaceful garden. It is there that the memorable scene took place, since immortalized by a great painter, and which will never be forgotten by all those who, whatever they may be told, cannot imagine that these anxious preoccupations as to the future are idle trifles. Seated near this window, and gazing heavenward, the mother and son, seeming to feel their parting near, were communing together of those hopes of another life then engrossing the minds of all. "They conversed," says St Augustine, "with ineffable sweetness, forgetful of the past, turned to the future, and with lips stretched toward that immortal spring where the weary soul finds refreshment. Separating themselves by degrees from the things of the body, and rising more and more in thought towards that life which endeth not, 'at length with a bound of the heart, they touched it.'" A few days after this conversation Monica died, and, in dying, gave the last and strongest proof of the change wrought in her by the ardour of her beliefs. Her son tells us that hitherto, like all persons of her time and country, she had been very anxious as to her sepulture. She had prepared a tomb near that of her husband, and her greatest consolation was to think that death would unite her again to him whose inseparable companion she had been during life. Yet, when she felt herself dying, she, of her own accord, renounced it. "You will bury your mother here," she told her children, and when asked whether she did not dread to leave her body so far from her country, she replied: "Nothing is far from God, and it is not to be feared that at the end of the ages He should be unable to find the place where He must raise me to life again." Augustine did as his mother requested, and buried the holy woman in one of the churches of Ostia.

In these days it requires a violent effort to re-awaken those great memories upon the silent shore. All is so changed, all seems so calm, so dead, that it is hard to picture the time when it was animated by the movement of life and the stir of business. And yet this solitude contained one of the most bustling towns of the world, and the place of this desert was filled by fertile fields. Where now only arid sands are seen, there used to be delightful shades and gardens growing luscious fruits. It is related that the Emperor Albinus, who passed for a refined epicure, held the melons of Ostia in high esteem. Pliny the Younger has vaunted the beauty of this shore, thronged with pleasure-houses as large as towns and rich as palaces; now one scarcely sees at far distances apart a few miserable cabins. In our days there is not a Roman who would consent to stay for one short hour after sundown upon these plague-stricken shores. We have just seen in the *Octavius* that, in the second century, people came hither from Rome to seek repose and health. The *isola sacra*, where a few scanty herds of buffaloes pasture, was one of the most lovely spots on earth, so full of verdure and flowers that it was looked on as one of the best-loved abodes of Venus. At Rome I have often heard it said that this ancient prosperity might return; that by better cultivation the country would be rendered healthy; that it would be easy to drive away the fever by draining off the stagnant waters, and that one might succeed in reclaiming a whole large useless territory. It seems to me that this ambition is of a nature to tempt Italy. The Italians have this good fortune, in addition to so many others, that, in order to spread, they have no need to attack their neighbours, and may make conquests without leaving home. They are quite right in affirming that they have not yet redeemed all their ancestral heritage; but this part of themselves of which they have not resumed possession, this *Italia irredenta* which absorbs and impassions them, is at home, in their country, at their doors. Near their great towns so fraught with life and beauty, they will find, if they choose, dead cities to revive. Instead of maintaining that martial state which exhausts them, and listening for the least sounds of external discord, in order to turn them to account, they may busy themselves with repeopling their deserts, in tilling their sterile lands—in short, in giving back to Italy all those rich domains which the negligence or barbarity of preceding ages has lost her. This is an enterprise that will expose them to no risks, and which the world will applaud.

CHAPTER VI.

POMPEII.

I.

THE EXCAVATIONS AT POMPEII UNDER SIGNOR FIORELLI—MEMENTOES OF ITS ANCIENT HISTORY THAT HAVE BEEN FOUND—WHAT REMAINS TO BE CLEARED—OUGHT THE WORKS THAT HAVE BEEN BEGUN TO BE CONTINUED?—RECENT DISCOVERIES—THE FRESCO OF ORPHEUS—ACCOUNT BOOKS OF THE BANKER JUCUNDUS—THE NEW "FULLONICA."

ALTHOUGH Pompeii has been much spoken about, much remains to be said of it. The excavations continue, and have not ceased to be fruitful. They have, since 1863, been directed by Signor Fiorelli, one of Italy's most distinguished archæologists, a piece of rare good fortune which has produced the most happy results. Persons who have not been to Pompeii for some years past will be struck to see the new aspect assumed by the ancient town. Not only does everything appear better ordered, and the work carried on in a more regular manner, but when one walks along the streets, enters the houses by the open doors, and goes over a quarter entirely cleared, one feels that the illusion has become easier and more complete, and that one may enter into the life of ancient times with even greater facility than was formerly the case. This progress is due to Signor Fiorelli, and to the resolution made by him to break with old routine and proceed by new methods. We must again repeat that the old-fashioned mode of excavation has been entirely discarded by him. The people who, on the 1st April 1748, began to turn over the ashes that for seventeen centuries had covered Pompeii, had only one object —the desire to find masterpieces to enrich the king's museums. Hence, the way in which the works were carried on, is easily explained. Excavation proceeded by chance, and at various places at once, according to the hope that existed of some piece of good fortune. If nothing was found, the excavation, after a little searching, was abandoned, and they went elsewhere. When the litter became troublesome it was carelessly thrown on the houses already laid bare, which were thus given back to the darkness whence they had just been drawn. As for those left in the light of day, no care was taken to preserve them.

The frescoes that had not been judged worthy of transference to the museum of Portici or of Naples remained exposed to the wind and the sun, which quickly effaced their colours. The destruction of the mosaics was completed by the feet of travellers and workmen: the walls became cracked, and ended by falling. A few men of knowledge and understanding, like the Abbé Barthélemy, indeed, complained of the deplorable manner in which the excavations were conducted; but since, after all, they produced masterpieces, and, thanks to them, the museum at Naples had become one of the richest in the world, the malcontents were not listened to. As a matter of fact, and despite the slight increase of care induced by course of time, this barbarous system lasted down to our own days.

Signor Fiorelli changed all this. In his reports he said, and repeated, that the chief interest of the excavations of Pompeii was Pompeii itself; that the discovery of works of art must only be considered as secondary to it; that they were seeking above all things to resuscitate an ancient Roman city which should give us back the life of former times; that, for its teachings to be complete, it must be seen entire and with its meanest old houses ; and that they sought to know, not only the dwellings of the rich, adorned with their elegant frescoes, and lined with their rich marbles, but also the abodes of the poor, with their common utensils and coarse caricatures. With this aim in view, all became important, and it was no longer allowable to neglect anything. So Signor Fiorelli decided, before pushing the works further, to go over those of his predecessors. Treading everywhere in their footsteps, he had the walls which threatened to fall propped up and supported, raised those that were already down, protected the frescoes and the mosaics, and, at the same time, busied himself with the definitive clearing of all that had been covered up again with rubbish, or had not been excavated. This was a tiresome and apparently unprofitable undertaking, for he was certain not to find much that was new in ground which had already been explored. But in order to know the city as a whole, it was necessary that everything should be cleared and brought to light. Signor Fiorelli therefore decided not to dazzle public opinion by the announcement of unexpected discoveries for some time to come,[1] and to prosecute in silence a task more useful than brilliant. It took him twelve years to accomplish this seemingly ungrateful work; but the result amply repaid him for all the time and labour spent upon it. He who formerly visited Pompeii was stopped at every moment by mountains of cinders and islands of rubbish which hindered locomotion, blocked the streets, and interrupted his walks. Even about the Forum and quite close to the theatre, houses remained which had not been excavated. These breaks have now disappeared. The discovered part of Pompeii is entire. It is all spread out before our eyes, with its smallest alleys, its meanest houses, and its most humble shops, and walking through it one may obtain a very true and complete idea of life in ancient times. It must be owned that such a result deserved purchasing with some years of obstinate toil.

This labour of patience and detail led Signor Fiorelli to make some curious discoveries, of which a word must

---

[1] Yet it must not be forgotten that it was Signor Fiorelli who thought of running plaster into the void left by the bodies of the Pompeians in decaying. On completion of the operation the plaster gives the image of the dead exactly. It is, in fact, conceivable that when the damp ash which was spread over Pompeii became cool, it should have preserved the forms of the objects which it covered over, like a mould. Thus it is that it has been found possible to assemble in the little museum, placed at the entrance of the town, several people who are reproduced just as they were when death struck them; some wrestling against it in despair, others yielding without resistance. It is a striking sight, and one of the greatest curiosities of Pompeii.

be said. Pompeii, at first sight, produces the effect of a new and improvised town. Everything in it seems to be of the same character and the same period. It is, in fact, known that after the earthquake of the year 63, which greatly damaged it, its reconstruction was commenced, and that it was very far advanced when, sixteen years later, the town was covered up by Vesuvius. It was the age of Nero, a terrible artist, who had a furious taste for building, who wished to renew everything, and who, it is said, set fire to Rome in order to have the pleasure of re-making it again, according to the fashion of the day. The manias of the master, even when he was called Nero, were the Empire's law, and since the Pompeians had to repair their city, the occasion was used to change and rejuvenate everything. The temples were enlarged, the old buildings adorned with new façades, the walls covered with stucco or incrusted with marbles, and pillars of tufa were replaced by columns of travertine. "In short," says M. Nissen, " they were about to quite modernise the town, as during the old *régime* they used to disfigure venerable cathedrals under pretence of repairing them, and as the second Empire rebuilt Paris and the old towns of France according to line."[1] It is these restorations which now especially strike visitors. Passing quickly along, they only perceive stucco or marble linings and the solemn façades hastily erected in the time of Nero, without having time to see that the new buildings have covered up old foundations, without destroying them. Signor Fiorelli, who looked at everything closely, reached those

---

[1] Nissen, *Pompeianischen Studien*, 360.

solid basements that survived the earthquake and resisted the eruption of Vesuvius. Beneath the town of the second century he finds at least two yet more ancient ones, whose history he traces out for us. The oldest goes back to the sixth century before the Christian era. At that time some families, come from no one knows where, took possession of the ground extending from the Sarnus to the sea. They enclosed this ground with walls formed of enormous blocks, taken from the neighbouring mountains, and placed one upon the other, without cement. In this space, too vast for them, the new inhabitants settled at their ease. Their houses, whose foundations still exist, consisted only of a covered court, around which were distributed the apartments. Each habitation was placed in the midst of a small plot of ground (*hæredium*) tilled by the family. The town, therefore, was not at that time an agglomeration of houses crowded one against the other, but an assemblage of families living on their lands under the shelter of a common wall.[1] Two centuries later, the Samnites came. They were an intelligent and civilised people, and soon allowed themselves to be won over to the arts of Greece. They built a real town, with very fine monuments, of which some still exist, together with the inscriptions placed on them by the magistrates. Pompeii then attained a high degree of wealth and culture. M. Nissen bids us remark that she imitated

---

[1] Signor Fiorelli has summarised his ideas on the first times of Pompeii and on its history in the introduction to his *Descrizione di Pompeii*. His opinions are discussed and completed by M. Nissen in his *Pompeianischen Studien* and in the *Pompeianischen Beiträge* of M. Mau.

the Greeks much more frankly than Rome ventured to do at the same period. Thus, she had a palestra, where her young men came to exercise themselves, like those of Sparta or of Athens; she had a stone theatre, while as yet the Romans only built wooden stages that did not survive the games given upon them; and she openly raised a temple to Isis, who was only officially admitted to Rome at the period of the Flavii. It was, then, at this point of time, and long before she had become Roman, that Greek civilisation planted itself so deeply there. The loans which the little town so willingly contracted with foreign lands did not prevent it from highly cherishing its independence, which it defended bravely against the Romans during the social war, and Scylla had great trouble to subdue it. When it had been conquered, he sent three cohorts of veterans with their families thither, who formed a colony bearing his name (*Colonia Cornelia*). Its prosperity did not suffer from the new *régime*, which it accepted with a good grace. Some years later, Cicero, eulogising Campania, said the towns there were so elegant, so rich, and so well built, that the inhabitants had a right to laugh at the poor old cities of Latium; and among the beautiful towns, of which the Latins had cause to be jealous, he placed Pompeii.[1]

Since the preliminary works of Signor Fiorelli have been finished, and we have a more exact and complete plan of the quarters excavated up to the present time, it has been possible to recognise better than was formerly the case, that the town is regularly built,

---

[1] Cic., *De lege agrar.*, II. 85. Note that he does not speak of Herculaneum.

and that, in general, its streets are well outlined and cut each other at right angles. It must not be thought that this regularity was introduced into Pompeii by the architects who rebuilt it after its first disaster. Signor Fiorelli thinks that it already existed in the primitive town. The old Italians who first settled on the banks of the Sarnus had a particular method of constructing their towns, and usually built them on the same plan. After forming the boundary, they traced two perpendicular lines, the one from north to south, called the *cardo*, the other from east to west, called the *decumanus*: these were the two main streets, on which, later on, the others came to be branched. At Pompeii the *decumanus* and the *cardo* are still visible, and their direction being settled, and it being certain that the regularity remarked in the discovered quarters was repeated in the others, it is possible, with the help of the past known to us, to form an idea of the past which we do not know. Thus, Signor Fiorelli could approximately imagine a sort of plan of the whole. According to the extent of the ground and the direction of the streets, he divides it into nine quarters, or, as the Romans used to term it, into nine regions. Of these nine regions three are entirely cleared, three entirely covered up, and of the three others only a small part is known. Striking a balance, therefore, the part of Pompeii still to be discovered, and whose disinterment is being vigorously pushed forward, amounts to rather more than half.[1]

But is it well to do so? Would it not be better,

---

[1] According to Signor Ruggiero, the entire surface of Pompeii must be about 662,000 square mètres. Of these, 264,424 have been cleared.

instead of entering upon fresh excavations, to stop and divert this vigorous effort of research to newer and richer grounds ? This is what Beulé maintained, with great energy, in one of the best books ever written by him.[1] Beulé was even more an artist than an archæologian. Obscure finds that only serve to solve some historical problem, and render the past more living, gave him much less pleasure than the discovery of the statues, the mosaics, and the fine friezes which charmed his delicate taste. Now, he remembered that every time they had dug under Portici, or in the depths that conceal Herculaneum, they had returned with admirable works of art. " It is there, then, that you must dig," said he; "you must concentrate your efforts and your resources upon these untouched ruins which promise so many treasures." And with the ardour which he was wont to display in the propagation of his opinions, he invited all the friends of the Arts and all the rich amateurs of Europe to unite in order to cover the expenses of these fruitful excavations.

Should this appeal be ever heard, and bankers and antiquaries bring Signor Fiorelli the wherewithal to resume the costly works at Herculaneum, I believe he will right willingly accept the generous offering, and be glad to direct part of his workmen to this side.[2]

---

[1] *Le drame du Vésuve.*

[2] Public attention has lately been recalled to the excavations at Herculaneum in a somewhat unforeseen manner. At Pompeii, a kind of commemoration was being celebrated of the catastrophe which took place in 79, that is to say, eighteen hundred years previously. On that occasion the excavation committee published a volume of notices and memoirs, entitled *Pompeii e la regione sotterrata dal Vesuvio*

But I doubt whether, even in such a case, he would be induced entirely to give up Pompeii—that is to say, perhaps modest, but certain and easy success, for difficulties and risks. Why, indeed, should he consent to this, and by what reason could the abandonment be justified? Pompeii, says Beulé, has yielded about all that can be expected of it. In this new town, rebuilt and decorated in sixteen years by the same artists, everything is alike. Supposing the excavations to be as fortunate in the future as in the past, only the same house will always be met with, built of the same materials, divided in the same manner, with its atrium and its peristyle, its rooms for the slaves and for the masters, its apartments private and public. He adds that even this house itself, studied so many times, this elegant house in which there was always the hope of coming upon some precious piece of furniture, will not be found again. The rich quarters, *i.e.* those surrounding the Forum and the theatres, have been ex-

---

nell' anno LXXIX. Among these memoirs is a very curious one by Signor Comparetti, concerning the villa of Herculaneum, where the famous Greek and Latin papyri were found, and which he thinks belonged to a rich Roman, L. Piso Cæsarinus, Cæsar's father-in-law. We know that this villa was full of marvellous works of art, and that the finest bronze busts which we admire in the Museum of Naples were discovered there. In another memoir, quickly following that of Signor Comparetti, Signor de Petra, studying the reports of the engineers who directed the excavations in 1750, proved that only part of the villa was then cleared, so that, by taking up the works again, we should have some chance of perhaps also reaping a rich harvest. It must be confessed that the hope of finding some bronze or marble statue, like the Drunken Faun or the Æschines, is tempting enough to induce the resumption of the excavations so unfortunately interrupted.

cavated, and there is now little chance of finding any houses but poor ones: is it worth while to spend time and money for the sake of tumbledown hovels?

Signor Fiorelli might reply that, after all, these hovels too have their interest. We know the rich classes of antiquity pretty well. It is especially of them that history tells us, acquainting us with their ways of thought and living. On the other hand, neither poets nor historians have busied themselves much with the poor. What a service Pompeii would render us by putting before our eyes a sort of living picture of the popular classes of the Empire! Thus, even if the certainty existed that only poor habitations remain, it would not be a reason for suspending the excavations. But M. Beulé's prediction has not been fulfilled. We have gone on finding as many elegant houses in the new quarters of Pompeii as in the old ones, and within the last few years discoveries have been made as curious as those of former times. No better answer can be made to those who are inclined to think Pompeii an exhausted mine, and who seem to believe that it can no longer repay us for our trouble, than to show them by some examples that the latest excavations there have not been less fortunate than the others.

Firstly, they have not left off finding interesting paintings. There is scarcely a house but contains one, and the catalogue drawn up by Signor Sogliano of all those discovered within twelve years [1] includes

---

[1] Signor Sogliano's memoir is contained in the work I have just spoken of, and which was published on the occasion of the commemoration of the destruction of Pompeii.

over eight hundred, some of which are very curious. Space being limited, I will only instance the fresco of Orpheus, not that it is more remarkable than the rest, but because a picture much resembling it was found in one of the Christian cemeteries at Rome. The two differ little, except as to their dimensions. That of Pompeii measures about $2\frac{1}{2}$ mètres. Consequently, the details are better marked and more visible than in the painting of the Catacombs, which is smaller, and has been much effaced by time, yet the general aspect of which is the same. Orpheus is represented sitting, a light *chlamys* falls from his shoulders over his limbs, he plays with the *plectrum* upon the nine-stringed lyre. At his feet the Pompeian painter has crowded very different kinds of animals—a lion, a tiger, a panther, a boar, a deer, and a hare. Further on, there are trees and rocks drawn by the charm of his voice, and a brook which stops in its course to listen the longer. The Christian artist has suppressed all these animals and replaced them by two sheep. He doubtless wished to recall the memory of the Good Shepherd, who was the usual, and, as it were, official image of Christ, in the first times of the Church. But as a whole, he had reproduced the pagan fresco. He could do so without scruple; that beautiful face, so grave, so sweet, which seems wholly engrossed with the subject of the songs, without perceiving the strange effects they produce, has in itself something religious; Christianity had nothing to change, in order to adapt it to its worship and its dogmas, so, as we have already seen, it did not hesitate to represent Christ under the same forms as the pagans had given to the singer of

Thrace. Comparison of the Pompeian Orpheus with that of the Catacombs shows manifestly with what facility the rising Church borrowed antique types, and the importance which must be given to the imitation of Greek models in the birth of Christian art.

We must linger somewhat longer over a very curious and unforeseen discovery, made in the house of the banker L. Cæcilius Jucundus. At first sight this house does not appear much better than others; on the contrary, it is built in a rather narrow street, and is of modest appearance. Jucundus does not care for exteriors, and perhaps even, like a prudent man, he was glad not to blazon his fortune too much. But on entering we soon perceive that we are in the house of a rich man. The reception-hall is adorned with mythological pictures, and in the peristyle a great hunt is depicted. This painting is not, however, the most curious thing found in the peristyle. In searching above the embrasure of a door they found, in a well-hidden spot, the Pompeian banker's account-books. This was a great novelty, for books seem to have been very rare in Pompeii; while at Herculaneum, where only a few houses are known, a library was discovered almost at once. During all the time (more than a century) that Pompeii has been in course of excavation, neither waxen tablets, rolls of papyrus, books of parchment, nor archives of any kind have been found. This is not easy to explain.[1] Doubtless, Pompeii was not a

---

[1] The most likely explanation is that the hot ash which covered up Pompeii consumed the papyri, whereas the torrent of mud that flowed over Herculaneum, and which rose to a height of 20 mètres above the town, preserved them.

seat of study, and the learned could not have been numerous there. Yet books are not out of place, even in pleasure towns. If one of our beautiful sea or thermal-spring residences, whither one does not go to study, were to be swallowed up by a cataclysm, I suppose that, on bringing it to light again, not many works of science, but a good collection of novels and newspapers, would be found. Supposing there were no philosophical books at Pompeii, as at Herculaneum, the poets who sang of love must at least have been read there, since their verses are everywhere pencilled on the walls, and it would seem as if one ought long since to have found some copies of the elegies of Propertius or of Ovid's *Art of Love*, but all has been lost. The only indication pointing to the conclusion that the Pompeians sometimes bought books, and that, consequently, they must have had them in their houses, is the sign of a bookseller's shop, near the Stabian gate, which appears to have been carried on by four partners. Unfortunately, if the shop remains, the books have disappeared. So the joy which was felt may be easily understood when, on 3rd July 1875, it was seen that not a real library, but what might be called the pocket-book of Banker Jucundus, had been discovered.

It was a rather large box, placed in a sort of niche above a door, and containing a large number of the tablets (*tabulæ*) on which the Romans wrote the rough copies of their business papers, their little unimportant notes, the first draft of the books they composed—in short, all their current writings, reserving parchment and papyrus for what they desired definitively to preserve.

These tablets usually consisted of two or three thin slips of wood, joined together like book covers, and spread over inside with a light coating of wax. This was written on with an iron stylet. Yet it is a thing so frail, so delicate, so little made to endure, that has survived accidents of every kind which marble and iron could scarcely resist. One asks oneself by what miracle —in the midst of a city submerged and engulfed under that rain of stones and ashes which covered all the houses—this wood and this wax have not been destroyed, and one is still more astonished that, after so terrible an adventure, they should have been able to pass through eighteen centuries of darkness and damp without finally perishing. As a matter of fact, the Pompeian tablets reached us in a very sorry plight. When found, nothing was left but an assemblage of calcined cinders, and scarcely were they touched by the rays of that sun which they had not seen for eighteen hundred years, when they were seen to split on every side and fall into crumbs on contact with the air. The transport of these precious remains to Naples had to be effected with infinite precautions. There, in those work-rooms where, with admirable patience, the unrolling and reading of the papyri of Herculaneum are carried on, they set to work to separate the tablets one from the other, to join the scattered fragments, to open them out, and, when fortunately the wax was not melted, to decipher the traces left in it by the iron stylet. Thanks to Signor de Petra, the clever and learned Director of the Museum at Naples, the success attained was greater than had been hoped for. He superintended the work, and, when it was accomplished, first made known its results to the public.[1]

Are these results commensurate with the trouble they cost? It must be remarked that discoveries of this kind have always been followed by disappointment. We begin by expecting too much, and very naturally the reality falls short of our hopes. After all, it was not to be supposed that a banker's house should contain many works of first-class literature, and it is not at all surprising that account-books should have been found there. The box of Jucundus contained 132 signed receipts, and of which 127 have been deciphered, wholly or in part. Almost all these receipts have reference to sales by auction, and complete our knowledge of its mechanism. Sales by auction (*auctio*), which now serve us for the purpose of getting rid of our books, our furniture, and our pictures, after having been reserved among the Romans for forced sales—that is to say, those which the State made of the goods of the condemned, or creditors of those of their debtors—had finished by being used for every other kind of sale. Signor de Petra remarks that this method of selling had become so general that the words *auctionari* or *auctionem facere* were regarded as mere synonyms of *vendere*. In important towns there were large halls with courts and porticoes expressly built

---

[1] Signor de Petra's memoir, entitled *Le tavolette cerate di Pompei*, first appeared in the collection of the Academy of the *Lincei*. Since then, M. Mommsen has studied the tablets, especially from the juridical point of view, in an important article in the *Hermes*, XII. p. 88. In France M. Caillemer has busied himself with it in the *Revue historique de droit français* (July 1877).

for sales of this kind, called *atria auctionaria*. He who presided at the sale—the chief auctioneer, as we should call him—had to know how to keep accounts and draw up a regular report, so a professional banker was often appointed to the office. This is how, at Pompeii, Cæcilius Jucundus came to be charged with it. The presidency of the banker had, besides, another advantage. When the buyer, who was obliged to settle at once, had not the needful sum at his disposal, the banker advanced it. So in transactions of this nature he made two kinds of profits—first, the commission levied on the total proceeds of the sale in payment of his trouble, and then the interest required of the buyer for the money lent to him. Our tablets which, with a few unimportant differences, are all written the same way, contain the receipt of the seller to the banker who furnishes the funds, and represents the real buyer of whom he is the intermediary. These documents have especial interest for lawyers. Others, unfortunately in too small numbers, give us curious information touching the finances of Roman municipalities, and the way in which they administered their properties. They are signed by the town treasurer, and show us that Cæcilius, who was not satisfied with the emolument accruing to him from sales by auction, also undertook to manage the communal estates. He had thus taken farm pastures, a field, and a fuller's shop belonging to the municipality, perhaps either sub-letting or working them himself. Such were the means hit upon by the banker of a small town in order to enrich himself. The receipts of Jucundus enable us to seize in its essence a profession of which we know but little. They do not, therefore, lack importance, but, above all, they have revived in the learned world the almost lost hope of one day finding among the ruins of Pompeii some library, or at least some archive a little richer and more lettered than that of the banker Jucundus.

Facing the banker's house, a *fullonica*, or fuller's shop, was brought to light. Several others were already known, and especially one celebrated on account of the interesting paintings that were found in it, representing in a very clever, spirited manner all the operations of the trade. This trade was then very important. In the capital and in the provinces all Roman citizens who respected themselves wore the toga. It was the garment of elegance, the official and ceremonial dress, which characterised and distinguished the masters of the world—

"*Romanos rerum dominos gentemque togatam.*"

But if the majestic fulness of the toga, the elegance of its folds, and the brilliancy of its whiteness, above all, when cast into relief by a purple band, made it one of the finest dresses ever worn by men, it had the double disadvantage of being inconvenient and of easily getting soiled. When it was desired that it should be cleaned, and do honour to him who was to wear it, they sent it to the fuller's. There they began by throwing it into tubs full of water, chalk, and other ingredients. Then it was washed, not as it is now done, with the hands, but by treading (*foulant*) it with the feet. The workmen charged with this task performed in the tub a sort of *trois temps* movement (*tripudium*), like that of the wine-maker pressing the grape. By a strange accident the *tripudium* became the religious and national dance of the ancient Romans. This it was which the brothers Arvales executed while singing that song to the Lares, preserved to us by a chance, or the Salii performed when, in the month of March, they passed through the streets of Rome, beating with their little swords upon their brazen shields. When the linen was thus washed, they spread it upon a wicker cage, where it was exposed to the fumes of sulphur. Then they stretched it, carded it with a long brush, and finally placed it under a press very like those used in the vintages. The more it was pressed, the whiter and brighter it came out.[1] These various operations required large premises and numerous workmen. So in ancient towns there were fullers in great number. They passed for jovial, pleasure-loving, jocular folk, and Roman popular comedy therefore loved much to busy itself with them, and to put them upon the stage. The sight of the "jolly fullers" (*fullones feriati*) was privileged to amuse the people. The discovery of the new *fullonica* proves to us that the fullers of Pompeii resembled those of Rome. On the wall of the portico where they washed the wool the remains of a large painting were found, unfortunately much effaced, but which appears to have been drawn with much comic power. It is thought to represent the

---

[1] In the new *fullonica*, the room which served the workmen as a workshop is in wonderful preservation. One would think work had only just been left off. The basins into which the washing was thrown are intact, and one would think that the iron taps, which have remained in their places, were about to pour the water of Sarnus into them. In a corner is seen an urn full of the chalky matter put into it the evening or the day before the eruption.

feast of Minerva (*quinquatrus*), also that of the fullers. In it persons are seen yielding themselves up to joy with such violence that their games sometimes end in blows, and one of them, who has been beaten until blood flowed, comes to appeal to the law. But gay scenes prevail. There are dances and feasts where the guests are depicted in grotesque or obscene attitudes which Rabelais alone would dare to describe. This freedom of brush reminds us that we are in the land where the *atellana* was created.

A thing to be noted is that the new *fullonica*, the house of Jucundus, and that which contained the Orpheus, are near each other. If we have been able to find so many curiosities nearly at the same time in a single corner of the town, must it not be concluded from this that we do well to continue the works, and that in prosecuting them still more happy discoveries may be expected?

II.

POMPEII'S CHIEF LESSON TO US — COUNTRY LIFE IN THE ROMAN EMPIRE — THE DIFFICULTY OF ACQUAINTING OURSELVES WITH IT — HOW POMPEII PUTS IT BEFORE OUR EYES — THE WHOLE EMPIRE REPEATS THE CUSTOMS OF ROME — THE ARISTOCRACY OF POMPEII — CHARACTERISTICS OF POMPEIAN HOUSES.

THESE new discoveries, added to those made in the course of the previous century and a half, certainly make Pompeii one of the most interesting places in the world. By a rare privilege one instructs as much as one amuses oneself, and this journey, which delights the curious, is a source of still greater pleasure to people who wish to learn. Now that nearly half the town has been cleared, and it has become so easy to go over it, it is reasonable to ask ourselves what particular kind of profit may be drawn from visiting it, and, above all, what it teaches serious minds who make it their study.

It appears to me that the great use of Pompeii for us is to teach us what provincial life was like in the Roman Empire. We know very well how time was passed at Rome, ancient authors being full of precise information on the subject. In Cicero's letters we can live the day of a statesman over again. Horace's satires paint for us to the very life the existence of a lounger whose chief occupations were to walk in the Forum or along the Sacred Way, to look at the ball-players in the Field of Mars, to chat with the corn or vegetable merchants, and in the evening to listen to the quacks and the fortune-tellers. Juvenal, more indiscreet, allows us a peep into the interior of a dreadful tavern, the trysting-place of sailers, robbers, and fugitive slaves, at the end of which the officials of the funereal pomps sleep side by side with the begging priests of the Great Goddess. What escapes us is provincial life.[1] We should probably know it better had the whole of the Latin theatre been preserved to us. Since the inhabitants of large towns rather like

---

[1] I here use the word *provincial* in the French sense for all that was not Roman, and, consequently, for Italy as well as for Gaul or Spain. The Romans made a distinction, and did not include Italy in what they called the *provinces*.

to joke at the absurdity of small ones, it may be supposed that the authors of the mimes and of the *atellanæ*[1] did not fail to turn them into ridicule. This is proved, too, by the titles of some of their pieces, and the short fragments we have preserved of them. Pomponius and Novius more than once amused themselves in describing the misadventures of a candidate. Doubtless the elections of some small municipality were in question, for the Romans would not have suffered those of Rome to be laughed at. In a piece entitled the "Setinian," the poet Titinius put upon the stage one of those inveterate provincial dames who picture to themselves the entire world as turning round the axis of their village. They bring everything into connection with it, and believe that all is made for it. This one, while they show her Rome, only thinks of her dear Setia. "Ah," she replies to those who point out the Tiber to her, "what a service would be rendered to the territory of Setia if it could be made to flow there!" Unfortunately, only very short fragments are extant. These pieces have been almost entirely lost, and the little that remains only excites our curiosity without satisfying it.

If we turn to the writers who have come to us entire, we are scarcely less fortunate. As a rule, they only speak to us of the provinces, in order to tell us the deep repugnance they feel for them. Among

---

[1] From Atella, a town in Oscan Campania, where they were invented. An *atellana* was originally a species of pantomimic farce, composed of tricks, grimaces, cuffs, and contortions. Its character was, however, greatly modified in the hands of Roman writers, whose *atellanæ* much resemble other contemporary comedies.—*Translator.*

the lettered and the witty, they were no more fashionable then than now; all agreed in declaring that it was not possible to live out of Rome. Doubtless they were obliged to own it to be one of the most unhealthy abiding-places in the world. Fever had its altars there, even from the days of Numa, nor had it been decreased by the prayers made from such ancient times downward. Seneca owns that, in order to feel better, it was enough to quit this heavy, dusty, smoky atmosphere for a moment—yet it was never willingly left. Cicero, while quietly living there in it, did not scruple to say, even in his public speeches, that it was a very ugly, ill-built city; that the houses there were too high and the streets too narrow.[1] Yet he changed his mind directly he was out of it. "How beautiful it is!" he exclaimed. It sufficed him to have been banished from it only a few months in order to find it admirable.[2] He left it, however, a few years later, in order to govern Cilicia, but this time he began regretting it as soon as it was out of sight. He was thinking of the means of returning thither, even before he had reached his province. While administering lands more vast than kingdoms, commanding armies, and receiving the compliments of the Senate for his victories, he could not console himself for being so far from the Capitol, and wrote to his friend Cœlius disconsolate letters, advising him never to leave Rome, and always to live in its light: *Urbem, urbem, mi Rufe, cole et in hac luce vive.*[3]

---

[1] Cicero, *De lege agr.*, II., 35.
[2] Id. *Post red., ad pop.*, 1.
[3] Id., *Ad fam.*, II., 12.

Strictly speaking, it is conceivable that a statesman should not care to lose sight of the Forum, since it was so much to his interest not to absent himself from it. What is more surprising is that even poor people, for whom life in Rome was so dear and so difficult, persisted in remaining there too. Juvenal has most eloquently described to what wretchedness a poor client like himself was every day exposed. In order to gather courage to leave it, he vaunts to himself life at Sora, at Fabrateria, and at Frusino, charming towns, where there is no danger of being crushed in the morning by vehicles, and murdered in the evening by robbers; where a house and a garden may be bought at the cost of the annual rent of a dark lodging in Rome. "Ah," he says to himself with an emotion at which we cannot help feeling touched, "it is there thou must live, enamoured of thy spade, and tending well thy little plot, which will bring thee vegetables enough to feast a hundred Pathagoricians. It is something to be an owner, no matter where, no matter in what corner, even of a lizard hole."[1] Yet Juvenal did not succeed in convincing himself. He remained at Rome, where Martial shows him to us in the morning, wearily climbing the slopes of the great and little Cælian, on his way to pay his court to the rich who protected him. Statius, at least, showed more resolution. He saw his reputation increase without his fortune augmenting. He was the first poet in Rome, yet one of the poorest, and, in order to live, he was forced to sing the *amours* of the rich and celebrate the

---

[1] Juvenal, III. 228.

virtues of Domitian in every key. What troubled him most was that he had a grown-up daughter to marry, a girl full of talent, who played upon the lyre and sang her father's verses ravishingly. Unhappily he had no dowry to give her, and "her beautiful youth was flowing away barren and lonely."[1] He resolved to return to Naples, his birth-place, where he hoped to find an easier existence and less exigent sons-in-law, but his wife refused to follow him. She was one of those obstinate Roman ladies who thought it impossible to live elsewhere than on one of the seven hills. At thought of leaving Rome she emitted deep sighs and passed sleepless nights. In vain did Statius describe to her, in delightful verses, the marvels of Puteoli and Baiæ, that enchanting country, "where all unites to lend life charm, where the summers are cool and the winters mild, where the sea comes peacefully to die upon those shores which it caresses." She only thought of Suburra and the Esquiliæ. She was a woman capable of regretting the brooks of Rome in presence of the sea of Naples.

This repugnance felt by Roman men of letters for the provinces explains their silence concerning them. One does not care to speak of things which displease one, so they mention it as little as they can help, and what they say teaches us nothing precise or new. We should therefore now be very much embarrassed to divine how life used to pass in a little town of the Roman Empire, if one had not fortunately been found. The discovery of Pompeii quite consoles us for the

---

[1] Statins, *Silvæ*, III. 5, 60.

silence of ancient writers. In order to know how people lived outside Rome, we need no longer with great trouble gather trivial and doubtful texts, for a short walk in Pompeii teaches us infinitely more.

We may make up our minds, before entering, not to find ourselves so far away from home as we might have been tempted to expect. Everywhere where an important capital exists, it exercises a sovereign attraction upon other towns; its monuments are imitated, its fashions are copied, its language is reproduced, its life is lived. In the first century the whole universe had its eyes turned towards Rome, whose usages had penetrated everywhere. Only Greek civilisation still resisted. The East defended itself energetically against what it called an invasion of barbarians, but in the West the most vigorous and rebellious nationalities had let themselves be subdued. Spain, Gaul, and Britain were subjected to the customs as well as to the laws of the victors. As our friends beyond the Rhine say, the world had become *Romanised*.

Roman influence insinuated itself into the most distant countries from several sides at once. While the legions, crossing the Empire on their way to their frontier camps, caused it to penetrate among the ranks of the populace, by that natural affinity which everywhere links the people with the soldiery, or even the traders who settled after the armies communicated, imposed their customs and their language upon the merchants, the agriculturists, and all those who had dealings with them, either for the sale of their own products or the purchase of those of Rome. As for the upper classes, they were in contact with the intendants (*procuratores*), the proprietors, and the proconsuls whom the Emperor and the Senate sent to govern the provinces. These were always people of the best society, knights or senators, accustomed to frequent the palace of Cæsar, and who brought, as it were, an air of Rome into these distant countries. They were often accompanied by their wives, and always brought with them sons of great families, who came to instruct themselves in business by their example, and freedmen who served them as secretaries. These formed a kind of court, on which the good society of the towns where they resided modelled itself. By this daily contact with merchants, soldiers, and governors, the provinces had become Roman. Tacitus says that people there carefully read the journals of Rome in order to keep themselves informed of the least events that passed in the Senate or the Forum;[1] they repeated the jokes against the masters of the moment, and were anxious to know the fine phrases and brilliant thoughts of renowned orators. The new works of fashionable authors were read everywhere. The librarians of Lyons advertised the latest pleadings of Pliny; those of Vienna sold Martial's epigrams, and this poet proudly tells us that his verses were sung as far as Roman sway extended. Even among the peoples little known and incompletely subdued, Rome penetrated as much by her arts and her literature as by her arms. "Gaul," says Juvenal, "has educated British lawyers, and it is said that Thule thinks of getting a professor of public eloquence."[2] Juvenal means to joke, but he does not

---

[1] Tacitus, *Ann.*, XVI. 22.   [2] Juvenal, XV. 110.

exaggerate so much as he thinks. Britain was one of the Empire's last conquests, and apparently one of the least stable; yet how great was its anguish when at the moment of the invasions it was forced to sever from Rome is well known. It is therefore probable that these distant provinces, these remote lands, held many a surprise in store for the Roman visiting them. He must have been greatly astonished not to feel himself so very much abroad. He even sometimes found there what is most difficult to transport from one country to another—that elegance of manners, that refinement of speech, that particular turn of raillery—in short, all those delicate qualities which the Romans understood by the word "urbanity," because they believed them bound up with life in the great city. When Martial came to Bilbilis, in the heart of Spain, he believed himself in a country of savages, and groaned at the thought of being there. But what was his surprise at finding a veritable Roman lady in the place! Even making allowance for politeness, his praises of Marcella show that *urbanity* had penetrated even as far as Bilbilis. "Speak but a single word," he tells her, "and the Palatine will think that thou belongest to it. None of the women born in Suburra, or who dwell on the slopes of the Capitol, can rival thee. Thou alone softenest my regrets at having left the mistress city. Alone thou sufficest to make it quite live again for me!"[1]

If the fine manners of the Capitol and the Palatine were found again in the heart of Spain, if rhetoric was

---

[1] Martial, XII. 21.

studied at Thule, if the customs, the fashions, and the manner of speaking and living of the Romans were faithfully reproduced at the ends of the world, it is clear that this imitation must have been much more visible in an Italian city, and, above all, at Pompeii—that is to say, at the gates of Baiæ and Naples, whither the elegant youth of Rome went every year "to enjoy the warm baths and the enchanting spectacle of the sea."[1] These distinguished visitors spread the habits of the great town around them, and the inhabitants of Pompeii could grow familiar with them while scarcely leaving home. This influence must have made itself felt by everybody, but it was especially the rich, those forming the country aristocracy, who had before their eyes models whom they gladly sought to reproduce.

There was in every age an important aristocracy at Pompeii, but that which governed the little town at the moment of its destruction does not appear to have been very ancient. It has been remarked that inscriptions anterior to the Empire contain names of magistrates which do not re-appear later on. The families of these personages seem subsequently to have vanished or to have become obscure. With the first Cæsars, the Holconii, the Pansæ, etc., appear in their stead. Are we to believe that the great events which then took place were not without bearing on their sudden fortune? Their generosity proves to us that they were very rich, and wealth only comes thus suddenly to skilful workmen, bold merchants, and fortunate speculators. Let

---

[1] *Propter aquas calidas deliciasque maris.* This is a line from an epitaph found at Ostia.

us not forget that Pompeii, which appears only to have been a pleasure resort, was also a town of commerce. Strabo affirms that it served as a port of Acerra, Nolæ, and of Nocera, so it was a kind of industrial centre for all this part of Campania. It is very possible that the impulse given to business by the establishment of the Empire, the peace and safety restored to the world after so many troubles, together with the increase of public well-being and opulence, which were their natural result, suddenly brought to the first rank families whose position had hitherto been more modest, and founded those great houses which were to predominate in the city for the ensuing century. That this aristocracy should have taken pleasure in imitating the manners of the Roman nobility whom it occasionally saw upon its shores is not to be wondered at, its position in the little town being nearly that of great personages at Rome. Like them, it filled all public offices, and, like them, too, it won and paid for the favours of the people by incredible liberalities. The two brothers Holconius rebuilt the whole of a theatre at their own expense. The inscriptions on monuments constructed by them, or which were raised in their honour, inform us concerning their public life; what their private existence was is less easy to ascertain. Until we have the good fortune to lay our hands on their account-books, as happened in the case of the banker Jucundus, some idea of their mode of life may best be gathered from the richness and beauty of their dwellings.

If we would appreciate the beautiful houses of Pompeii as they deserve, and properly understand the charms they must have had for their owners, we must lay aside a few prejudices. The dwellers in this charming town appear above all things to have anxiously sought their well-being, but they did not place it where we do. Each century possesses its opinions and its preferences in this respect, and the way to be happy has its fashions like everything else. Were we to allow ourselves to be dominated by that tyranny of habit which forbids us to believe it possible to live otherwise than we do, the houses of Pompeii would perhaps appear small and inconvenient to us, but if we for a moment forget our ideas and our habits, and make ourselves Romans in thought, we shall find that those who lived in them made them very well for themselves, and that they were perfectly adapted to all their tastes and all their wants. In our great towns it is now difficult, even for the rich, to possess a house for themselves alone. They mostly lodge in houses which they share with many others. Their apartments are composed of a series of large airy rooms, with broad windows that take in air and light from the streets and squares. There is nothing similar at Pompeii. The number of houses there occupied by single families is very considerable. The chief rooms are all on the ground floor.[1] The richest have built themselves houses situated between four

---

[1] The upper stories must have been reserved for less important rooms. They are reached by steep and narrow stairs. There is nothing like the main staircase of modern houses, which gives admittance to all the stories, and is common to every apartment. In Nissen (*Pompeian. Stud.*, p. 602) will be found some very astute observations on the part played in our dwellings by the staircase, and the character it has given them. Of all the parts of a modern house, it is that which a Pompeian would have least understood.

streets, and occupying what was called an entire "island." If chary of their fortune, they parcel out a portion of this great space for shops, to be let out at high rents. These shops sometimes take up all the outside frontage of a dwelling. While we carefully reserve the façade for the finest apartments, at Pompeii it is given up to trade, or shut in by thick walls without openings. The entire house, instead of looking on the street, is faced towards the interior. It only communicates with the outside by the entrance door, rigorously shut and guarded. There are but few windows, and these only in the upper stories. They like to live at home, far from the indifferent and from strangers. Now, what we term domestic life belongs in great measure to the public. The world has easy access to us, and when it does not come, we desire at least to see it through our widely-opened windows. Among the ancients, private life was more retired, more truly secluded, than with us. The master of the house did not care to look into the streets, and, above all, he did not want the street to look into him. He has even divisions and distinctions in his house itself. The part where he receives strangers is not that whither he retires with his family, nor is it easy to penetrate this sanctuary, divided from the rest by corridors, shut off by doors and hangings, and guarded by door-keepers. The master receives when he will, he shuts himself up when he chooses, and if in the vestibule some client more than usually troublesome or persistent awaits his coming forth, he has a back door (*posticum*) on a narrow street that allows him to escape.

To those who find the rooms of the Pompeian houses somewhat too small for their taste, it has already been replied that the inhabitants passed a great portion of their day from home, under the portico of the Forum or of the theatres. It may be added that if the chambers are not large, they are numerous. The Roman does with his rooms as with his slaves; he has different rooms for all the incidents of the day, as he possesses servants for all the various necessities of life. With him, each room is made exactly for the use it is to be put to. He is not, like us, satisfied with a single dining-room; he has some of different sizes, changing them according to the season and the number of friends he chooses to regale. The room where he takes his *siesta* in the day-time, and that where he retires at night to sleep, are very small, and only receive light and air by means of the door. This is not an inconvenience in the south, where shade brings coolness. Besides, he only stays there just as long as he sleeps. For the remainder of the time he has a court shut in, or nearly so, called the atrium, together with an open court or peristyle. Here it is that he likes to remain best when at home. He is there not only with his wife and children, but in view of his servants, and sometimes in their society. In spite of his taste for retirement and isolation, of which I have spoken, he does not shun their company. This is because the family of antiquity is more extended than ours, including in a lower degree the slave and the freedman, so that the master, in living with them, still considers himself among his own. These open and closed courts where the family pass their lives are found in all Pompeian houses without exception. They are indispensable for the purpose of affording light to all the rest of the premises. So even among the less rich, pleasure is taken in ornamenting them tastefully, and sometimes profusely. If the ground allows of it, a few shrubs are planted there, or some flowers are grown. Moralists [1] and men of the world laugh at these miniature gardens between four walls. Those who own magnificent villas, with large trees and arbours hung upon elegant columns, discourse of them very much at their ease. Each does as he is able, and I own that I could not be severe on those poor people who were determined to put a little greenery before their eyes. I am more inclined to find fault with them on account of their love for those brooklets which they pompously dubbed *euripus*, or for those grottoes of rock or shell work which are only so many pretentious toys. Their excuse is that this odd taste has been shared by the citizen classes of all countries and of all times. Those of Pompeii, at least, excel all others in their determination not to look on any displeasing object. They possess fine mosaics, brilliant stuccoes, and incrustations of marble on which their eyes may repose with pleasure. The fatiguing glare of the white stones has everywhere been softened by pleasing tints. The walls are painted grey or black, the columns toned with yellow or red. Along the cornices run graceful arabesques, composed of intertwining flowers, mingled at times with birds that never existed or landscapes which have nowhere been seen. These aimless fancies please the eye and do not exercise the mind. Now and then, upon some larger panel, a

---

[1] See what Fabianus says on this head (Seneca rhetor., *Controv.*, II. pref.).

mythological scene, painted unpretentiously and in broad strokes, recalls to the master some *chef d'œuvre* of antique art, and enables him to enjoy it from memory. Occasionally this humble citizen is so fortunate as to possess a bronze imitation of one of the finest works of Greek sculpture—a dancing satyr, a fighting athletic, a god, a goddess, a cithern player, etc.[1] He knows its worth, he understands its beauty, and has placed it on a pedestal in his atrium or his peristyle, to greet it with a look each time he passes out or enters in. Those rich Pompeians were happy folk. They knew how to beautify their lives with all the charms of well-being, to quicken it by the enjoyment of the arts, and I believe that many important personages of our largest towns would be tempted to envy the fate of the obscure citizens of this little municipality.

---

[1] It is from Pompeii and Herculaneum—that is to say, from two towns of the second order—that the fine bronzes of the Naples Museum come, which are the admiration of strangers. Among the citizens of our provincial towns nothing of the kind would be found. It must be added that the most beautiful things in Pompeii did not remain there. We know that after the catastrophe the inhabitants came and excavated for the purpose of carrying away their most precious things. Now, therefore, we have only what they could not find, or neglected to take.

III.

THE PAINTINGS OF POMPEII ACCORDING TO DOCTOR HELBIG'S WORKS — THE LARGE NUMBER OF MYTHOLOGICAL PICTURES — CHARACTER OF THESE PICTURES — THE PAINTINGS OF POMPEII NOT ORIGINAL—WHY CRITICS OF THE FIRST CENTURY TREAT THE PAINTINGS OF THEIR TIME SO SEVERELY—FROM WHAT SCHOOLS DID POMPEIAN ARTISTS BORROW THE SUBJECTS OF THEIR PICTURES—ALEXANDRIAN OR HELLENISTIC PAINTING—ROOM PICTURES—GENERAL CHARACTER OF HELLENISTIC PAINTING—HOW FAR DID POMPEIAN ARTISTS FAITHFULLY REPRODUCE THEIR MODELS—WHAT IS THE PARTICULAR MERIT OF THE PAINTINGS AT POMPEII?

WHAT seems to us especially worthy of envy in these charming houses are the paintings which cover nearly all their walls. They are the surprise and admiration of all who visit Pompeii. But it is not enough to look at them in passing, as is usually done. If we would carry away anything more than a fugitive impression of them, we must inquire of those who have made them their special occupation, and whom anterior studies prepared to understand them. By taking an enlightened connoisseur for our guide, we shall learn to appreciate them better, we shall have a more complete comprehension of them, and we shall succeed in acquiring from them a few correct ideas concerning the character and history of ancient art.

Professor W. Helbig is just one of those critics whose competence no one disputes, and who may be safely trusted. No one has studied the paintings of Herculaneum and Pompeii more thoroughly than he, and he has written two learned works about them, one of which forms the sequel to the other. The first gives us a minutely detailed catalogue of these paintings, with descriptions as precise as possible, and classifies them according to their subjects whenever these have fortunately been discovered.[1] In the other, the author handles all the questions raised by these paintings, seeking, above all, to ascertain how far the subjects are original, and whether the school to which they belong may be known.[2]

Of these two books the second is naturally the most pleasant reading, but the first, although drier in appearance, is perhaps of still greater utility. Even apart from the other work, which serves it as a commentary, this catalogue is full of the most curious instruction. I think that an age may be judged, not only by the books which it delights to read, but also by the pictures it most loves to look at. This is an indication as to its character and its tastes that can hardly deceive. Let us apply this rule to Professor Helbig's catalogue. Of 1968 paintings classified and described by him, there are rather more than 1400—nearly three-quarters—in some way or another connected with mythology,—that is to say, they represent the adventures of the gods or the legends of the heroic age. This figure shows the place held by the religious memories of the past in the minds of all during the first century. Even the incredulous and the indifferent felt their influence. When

---

[1] *Wandgemälde der vom Vesuv verschütteten Städte Campaniens*, Leipzig.
[2] *Untersuchungen über die campanische Wandmalerei*, Leipzig.

conscience escaped them, they still held sway over the imagination. This is a reflection which has often forced itself on students of the art or the literature of this period, but nowhere does it strike one more forcibly than at Pompeii. It is necessary to insist on this, if we reflect that, at the very moment when artists were profusely decorating Campanian towns with the images of gods and heroes, Christianity was beginning to spread in the Empire. St Paul had just passed quite close to these shores, on his way from Puteoli to Rome, and there are reasons to believe that the coquettish and voluptuous town about to be engulfed by Vesuvius had been visited by some Christians.[1] They preached their doctrine and celebrated their mysteries in these houses whose walls at every moment reminded them of a hostile worship. The multitude of these paintings gives us an idea of the obstacles which Christianity had to overcome. The religion against which it struggled had taken possession of men's entire existence. It was very difficult for a pagan to forget his gods; he came upon them everywhere, not only in the temples and public places of resort, which were filled with their images, but in his private dwelling, on the walls of those halls and chambers where he lived with his family, so that they seemed bound up with all the acts of his private life, and he who abandoned them appeared at the same time to break with all the memories and affections of the past. It was on these pictures that the child's first glances rested; he admired ere he understood them;

A word, thought to be *Christianus*, was found written with charcoal on a white wall (*Corp. insc. lat.*, IV. 679).

they entered his memory and became mingled with those youthful impressions which are never forgotten. The Fathers of the Church are therefore right in remarking that what then gave so many partisans to mythology was that it took possession of everybody in their cradles, and almost before their birth. So Turtullian said with as much vigour as truth: *Omnes idololatria obstetrice nascimur.*

Here we are then well informed, by the sight presented by the paintings of Pompeii, of the importance which mythology still retained, if not in the beliefs, at least in the habits of life. But what was the character of this mythology? How were these gods and heroes presented to their adorers, and in what adventures? Here again Professor Helbig's catalogue is very instructive. He shows us that what the painters prefer above all else are love stories. Jupiter appears engaged in seducing Danæ, Io, or Leda, and in carrying off Europa. The pursuit of Daphne by Apollo is the subject of twelve pictures, while Venus is represented fifteen times in the arms of Mars, and sixteen times with handsome Adonis. It is the same with the other divinities, and in all these pictures little else is thought of but their gallantries. This is what an elegant and trivial world had made out of an ancient and grave mythology. It is true that it had not offered much resistance. One of the greatest strengths of those ancient religions which possessed no sacred books, and were not fixed and held together by doctrines, was their ability to accommodate themselves easily to the opinions and tastes of each successive age. That of Greece sufficed for everything during centuries, and this is why it lived so long. From Homer down to the Neo-Platonians, it was able to take all forms—serious at times, at others sportive, always poetic—it served the poets for the purpose of expressing their most varied ideas and their most contrary sentiments, and allowed philosophers to clothe their most profound doctrines in brilliant colours. At the moment with which we are busied, it bent, with its usual suppleness and fertility of resource, to the caprices of a society fond of repose and pleasure, rich, happy, and assured of the morrow by a dreaded power, freed from serious political cares, and only left with that of passing life gaily, and which loved to represent itself under the image of its gods, and idealise their pleasures by attributing them to the inhabitants of Olympus. Thus we find another attraction in the paintings of Pompeii, when we reflect that they are the image of an epoch, and aid us to understand it. But since I just now spoke of Christianity, and showed that the strong affection remaining for mythology must be an obstacle to its progress, it should be added that it might render this obstacle less serious by showing what mythology had become, and that it was now nothing more than a school of immorality. It may well be thought that this was not neglected. Learned critics of our days have accused the Fathers of the Church of ignorance or calumny when they laugh at the gods and affirm that all the adventures attributed to them are only the glorification of the most shameful passions of men. They reply that these fables have a deeper sense, that they cloak great truths, and that they are, in reality, only an allegorical explanation of the most important phenomena of nature.

They are doubtless right, provided we only bear the mythology of primitive ages in mind; but it is certain that the mythology of the first century, at least in the minds of people of the world, had no longer this character. People who had the loves of Jupiter for Danæ or Ganymede painted in their houses were not sages desirous of expressing some cosmogonic thought; they were voluptuaries who wished to incite to pleasure or rejoice their eyes with an agreeable image. There is no longer the slightest intention of myth or allegory in them; only human life is represented, and the painter's thought does not go beyond the reproduction of love-scenes for the greater delight of the amorous. So when the Christian doctors so violently attacked the immorality of mythology, it was not possible to refute them, and they who listened to their invectives only had to raise their eyes to the walls of their houses in order to convince themselves that at the bottom they were not wrong.

The other paintings are either reproductions of animals and still life, landscapes, or character paintings.[1] These latter have a great interest for us, and render us many services. It is they which are looked at with the greatest curiosity in going over

---

[1] Among the character paintings, Professor Helbig distinguishes two different classes. There are first those in which a certain mixture of the real and the ideal is remarked, which, for example, represent Eros hunting, Cupids angling or grape-gathering, women engaged in their toilet with little Loves who help them, etc., and those quite realistic, which reproduce scenes of ordinary Pompeian life, without seeking to embellish them. It is these latter, especially, that will be meant when I talk of character pictures.

Pompeii. Representing real scenes and living personages, they seem to animate the desert town and give it back the inhabitants it has lost. But none of these various classes into which the Pompeian paintings may be divided can be compared, either as regards the talent of the painters or the number of the pictures, with that which contains only mythological subjects.

The first and most important question one asks oneself concerning the pictures at Pompeii is that of their origin. Whence came the painters who did them? Were they original artists who invented the subjects of their works? And if they were only imitators, to what school did their originals belong, and in what century did they live? As ancient writers give us no information on this point, we are forced to interrogate the paintings themselves, and draw all our knowledge from them.

With regard to the character pictures of which I have just spoken, the question is easy to solve. They represent local scenes and personages of the country, and were therefore created in the land itself, and taken from reality. If the master of the house which the artist had to decorate was one of those mad lovers of the amphitheatre or the circus, who wished to have their presentments always before his eyes, or if he merely liked everyday scenes, the artist pleased him by copying them exactly. He went to see the gladiators go through their exercises in the great barracks that have been found near the theatre, and reproduced them as he had seen them, and without more ceremony he transferred to his pictures the personages who frequented the Forum or the streets of the little town.

We may be assured that the fullers, the tavern-keepers, the bakers, and the fishmongers, seen on the walls of Pompeian houses, lived in the shops where their tools are still found. These half-naked women, with their hair raised above their brows in so strange a manner, are the same who sold their favours at a very low price in those narrow cells, which not everybody is allowed to visit, and which contain such gross drawings and brutal inscriptions. The painter had himself watched these peasants and workmen, in their monk-like hooded tunics, seated at a table before a glass of wine, whom he has rendered in a manner so true to life; he had seen with his own eyes this soldier, with his tawny complexion, full boots, and ample garment, saying gaily to mine host, who offers him a glass: "Come, a little fresh water" (*Da fridam pusillum*). What proves that the artist reproduced people of the country is, that they still strike by their resemblance to the people we have met on the piazza and in the shops of Naples. So the origin of these pictures is easily found. The artists who composed them faithfully imitated what they had before their eyes—they were done at Pompeii itself, and for Pompeii. But it must be remarked that they were very few (a score, at most) and generally of very slight dimensions. For the others, it is a different question. It does not seem to me possible to suppose the 1480 mythological pictures, which are often large works, and reveal a high talent for composition, are the work of original artists who painted them expressly to adorn the houses where they are now seen. Herculaneum and Pompeii were small towns, and scarcely deserved that a painter should go to so much inventive trouble for them. Besides, what proves that these pictures were not destined solely for them, is that they have been found in other places. Elsewhere, and especially at Rome, the remains of dwellings have been discovered, decorated exactly like those of the towns of the Campania.[1] The walls of these houses are painted with character pictures resembling those we admire in the museum at Naples, and with the same mythological subjects treated in the same manner: for example, Io, guarded by Argus and delivered by Mercury, seen in the house of Livia, in the palace of the Cæsars, exactly resembles six or seven compositions representing the same adventure at Pompeii. Does this not prove that these artists had prepared a certain number of pictures beforehand, had practised painting them, and reproduced them on every occasion when their services were required? But they were not really the creators of these pictures at Rome

---

[1] While digging along the Tiber shore, in order to widen the bed of the river, they found, opposite the Farnese Gardens, the remains of a charming Roman dwelling. It was composed of long corridors and a few rooms, of which one especially had been very remarkably decorated. When freed from the wet mud that had filled it for perhaps eighteen centuries past, the colours had an extraordinary brilliancy. As usual, architectural motives were observed, painted with great elegance, figures very boldly drawn, columns bound together with garlands and arabesques, and, in the midst, medallions containing scenes from common life—repasts, concerts, and sacrifices. This system of decoration is, on the whole, like that of the Pompeian houses, only more careful, and treated by artists of greater skill. These fine paintings, threatened with being again covered by the Tiber, were carefully removed and placed in the museum of the Lungara.

any more than at Pompeii. They neither imagined the subjects nor their arrangement. What allows us to affirm this is, that in scenes of any importance the conception is always superior to the execution. It exhibits a power of invention, a skill in composition— a talent, in fine, which seems above that of the obscure artist who did the fresco. It is, I think, natural to conclude from this that the painting was not executed by the same person who imagined the subject, and that the Pompeian artists, instead of taking the trouble to invent, were for the most part content to reproduce known paintings, adapting them to the places for which they were intended. The rapidity and inexhaustible fertility of their work are thus explained. Having in their memories, and, so to speak, at the end of their brushes, a crowd of brilliant subjects taken from illustrious masters, they could finish off the decoration of a house expeditiously and cheaply. They did not, therefore, paint from inspiration, but from memory; they were not inventors, but copyists.

This is probably why connoisseurs and critics of the first century treat the painting of their time so severely. We have on this subject the opinion of a clever man, an enlightened lover of letters and of arts, a person strange and full of contrasts, very light in his conduct and very serious in his judgments, who lived like the people of his period, and affected to think like those of former times. Petronius, in his satirical romance, imagines his heroes genuine adventurers, walking one day beneath a portico ornamented as usual with rare paintings. They look at them with great pleasure, desire to know their date, seek to understand their subject, and enter into a discussion. The past, as generally happens, swiftly brings them back to the present, and they soon begin to converse about contemporary art. They speak of it very severely, and the admiration they feel for the ancient artists renders them very hard towards those of their own age. They find that the arts are in full decadence, and that it is the love of money which has ruined them. In this connection follow the laments which have since then so often been repeated. "The past was the golden age, the arts shone in them with all their glory, because bare merit alone was loved. It is not surprising that they should have fallen off, when gods and men are seen by far to prefer an ingot of gold to all the statues and all the pictures which those poor Greeks, those madmen, Phidias and Apelles, took the trouble to make."[1] The conclusion is, that painting is dead, and that not even a trace of it remains. This opinion is almost that of Pliny the Elder, a less prejudiced and, in general, a more equitable judge. He somewhere asserts "that painting is about to perish," and in another place, "that it has already ceased to exist."[2] These are very rigorous verdicts, and people who have just visited Pompeii have some difficulty in subscribing to them. When they recall those scenes so skilfully composed, those figures so elegant and so graceful, and when they consider that these pictures were executed in so short a time by unknown artists for country towns, they find it impossible to believe that art was in such a desperate state as Pliny and Petronius affirm. But all is accounted for if we remember that those

---

[1] Petronius, *Sat.*, 2 and 88. [2] Pliny, XXXV. 29 and 50.

charming copies were, after all, but copies. They have not the merit of invention, and it is in invention that Pliny and Petronius, who pride themselves on being classic, made the greatness of painting chiefly consist. Since it no longer itself creates, and only lives by imitation, they consider it dead. Hence their severity.

We are no longer in the same position as they were. Since the originals have now ceased to exist, they cannot by comparison depreciate the merit of the imitations that were made of them. We no longer descend from the originals to the copies, which is always very dangerous for them; it is, on the contrary, the copies which enable us to soar to the lost originals, and picture to ourselves what they must have been. This service which they render us disposes us favourably towards them at starting. Far from complaining that the Pompeian painters were not inventive geniuses, we are almost tempted to be grateful to them for not having drawn upon themselves for anything. By being content to reproduce the creations of others, they carry us back towards those great epochs of ancient art which without them would be unknown to us.

But what, exactly, was the age in which the Pompeian artists sought these originals, and is it possible to ascertain with precision to what epoch of history and to what period of art the masters belonged from whom they drew their inspiration?

First of all, did they confine themselves to copying the pictures of a single school, and were they not of those eclectics who seek their advantage everywhere, reproducing the works of all times? They must doubtlessly have sometimes done so. Works are found among them differing from the rest, and which do not seem to belong to their usual manner. Such for example is the famous picture of the "Sacrifice of Iphigenia," one of the finest discovered at Pompeii, and which by a rare good fortune also happens to be one of the best preserved. In the centre, Iphigenia, in tears and with hands raised heavenwards, is brought to the altar by Ulysses and Diomedes. At one extremity Agamemnon covers his face in order not to see his daughter's death, and at the other Calchas, grasping the knife, seems sadly to prepare himself for his cruel part of sacrificer. Above, Diana appears in a light cloud, with the hind that is to be offered in place of the young girl. It appears to Professor Helbig, an expert judge in such a matter, that the regular arrangement of the picture, the symmetrical correspondence of the personages, the colouring of the background, and the folds of the garments, recall a somewhat ancient epoch of art. He draws attention to the circumstance that the figures are so arranged that the picture might be made into a bas-relief with scarcely any trouble. What is more characteristic yet is that Diomedes and Ulysses are represented smaller than Agamemnon and Calchas, in accordance with that somewhat naïve antique rule that the importance of the personages must be recognised by their height. While making all these curious observations, Professor Helbig does not go so far as to pretend that this fine picture belongs to a very remote age. In all times there are artists who love to turn back and resume old methods and processes. Pliny, speaking of two famous painters who worked at the temple of Honour and of Virtue, which

Vespasian was having rebuilt, says of one of them that he more resembled the ancients: *Priscus antiquis similior*.[1] The author of the "Sacrifice of Iphigenia" was doubtless an artist of this kind. Being a lover of archaism, he conceived and executed his picture in the antique manner, and the Pompeian artists, in accordance with their custom, faithfully reproduced it.

But these archaic fancies are rare at Pompeii. On the contrary, nearly all the pictures are very like each other, the subjects are usually conceived and carried out in the same way, and they appear to belong to the same school. Professor Helbig has no trouble in settling that it was the one which flourished at the court of the successors of Alexander. It was therefore Alexandrian or *Hellenistic*[2] art which the Pompeian artists imitated, and of which their paintings may afford us some image.

Although Greece was then in decadence, taste for the arts had not ceased to be as lively there as formerly. Alexander had honoured himself with the friendship of Lysippus and of Apelles; his successors, continuing the tradition, loved to gather artists around them, and sometimes became artists themselves. Attalus III., the last king of Pergamus, modelled in wax and chiselled bronze. Antiochus Epiphanes rested from the fatigues of royalty in a sculptor's studio. They grudged no-

---

[1] Pliny, XXXV. 120.

[2] German critics call *Hellenic* the literature which flourished prior to Alexander, and *Hellenistic* that which came afterwards. This designation is more correct than that of *Alexandrian literature*, for, under the successors of Alexander, there were very brilliant literary schools at Pergamus and Antioch, as well as at Alexandria.

thing in order to obtain statues or pictures that had charmed them. They paid artists insane sums. One of these princes proposed to the Cnidians, who were head over ears in debt, to undertake all their liabilities on condition of their ceding him the "Venus Aphrodite" of Praxiteles. Another, in the sale which Mummius made of the booty of Corinth, ran up the "Bacchus" of Aristides to the price of 100 talents (500,000 francs). Mummius, who could scarcely believe his ears, judged that a picture which people were willing to pay so dearly for must be a marvel, and kept it for Rome. The mad passions of these crowned amateurs knew neither bounds nor obstacles. Nothing was sacred in their eyes when it was a question of getting hold of a fine work. It is they who taught the Roman proconsuls the way to form a rich gallery for themselves at the expense of the most respected divinities. They were, in reality, the teachers of Verres. In their constant wars with each other, the treasures of the gods were no safer than those of kings. When Prusias I. invaded the territory of Pergamus, he did not in the least scruple to take from a venerated sanctuary the statue of Vulcan, a renowned work of Phyromacus. Ptolemy Evergetes, in his Asiatic expedition, under pretence of retaking the sacred images which Cambyses had carried off from Egypt, broke into the temples and seized all the works of art contained in them. Thus it came about that so many masterpieces were heaped together in the palaces of Pergamus, Antioch, and Alexandria. They were, however, not fated to remain there, for the Roman generals, taught by the example of the Greek kings, in their turn laid hands upon this rich booty, and carried it off to Rome, to adorn their triumphs.

From princes and kings, this taste soon descended to private individuals. The succession of Alexander, as we know, was the cause of trouble and wars without end. Never was power disputed with greater ardour, never more easily won and sooner lost than then. In such unquiet times great fortunes are quickly made and unmade. So these upstarts who remembered yesterday and feared to-morrow made haste to enjoy their ephemeral wealth. Menander's " Comic Muse " has popularised the type of those soldiers of fortune who came to squander in a few days with Athenian courtesans the money which they had gained in the courts of the sovereigns of the West. They love to show them, welcomed by their mistresses and flattered by their parasites as long as their golden Darics or Philips last, and then discarded and mocked, when they have nothing more in their purses. Among these parvenus were some who turned their fortunes to a better use. They imitated their masters by buying pictures and statues to adorn their houses.

It was a novelty: Professor Helbig thinks that, in the great age of art, artists scarcely worked for private individuals. Doubtless we are told that Agatharchus decorated the house of Alcibiades; but Alcibiades could not pass for an ordinary citizen. Painters usually kept their talents for the public. They covered the spacious walls of the porticoes with scenes taken from the ancient legends and the poems of Homer, or they composed pictures to be placed in the temples. Perhaps they may have thought that to make art subserve the pleasure of a single man would have been to degrade it. Pliny, at least, implies this, and he adds in magnificent terms that their pictures, instead of being shut up in a house, where at most a few privileged persons could enter, had the whole town for their dwelling; that everybody could contemplate them, and that a painter then belonged to all the universe: *Pictor res communis terrarum erat*.[1] But it appears that when, under Alexander, the Greek cities lost their freedom, their inhabitants became in some degree estranged from them. They felt less bound towards the Republic, since it no longer gave the citizens the same right, and they intervened less directly in its affairs. They were less proud of it, no longer cared so much to beautify it; in short, they thought less of it and more of themselves. They kept the money which was no longer destined for public monuments to decorate their own houses, which they made the centre of their existence. Painters naturally flattered a new taste certain to redound to their profit. "Two chief points," says Letronne,[2] "may be distinguished in the history of Greek art; that during which it was exclusively consecrated to keeping alive religious faith by images of the gods and paintings of their benefits, to awakening the patriotism of citizens by the ever-living spectacle of the great deeds of their ancestors, where, consequently, each work of the artist had its destination and its place marked in advance; and that in which art, so to speak, could no longer be had except by ordering, in which its productions became objects of luxury, were put in the rank of rarities, assimilated to industrial products,

---

[1] Pliny, XXXV. 118.   [2] In his *Lettres d'un antiquaire à un artiste*,

sought after less for their beauty than their price, and heaped together in the palaces of kings and of the wealthy for the vain pleasure of the eyes." Thenceforth the artist lost his taste for those large paintings which were made for a given monument, which had to correspond to the destination and architecture of the edifice, and which reproduced the character of the place they filled, and are only understood there. He worked in his studio according to his caprice upon subjects of his choice, without troubling himself as to what would become of his pictures, or, rather, certain in advance that a rich amateur would be found, ready to pay for them dearly and make them the ornament of his dwelling. Thus it was that, instead of large frescoes and vast canvases destined for public monuments, they began to paint what Professor Helbig rightly calls "room pictures" (*Cabinetsbilder*) in the same way that we say "chamber music," as opposed to that of the theatre or the church. They were to be hung along the walls in private houses, and became a kind of want, or, as it were, an indispensable luxury for those who were called the happy ones of the world.[1]

Professor Helbig has very well shown, in what is perhaps the best part of his book, that the Pompeian system of decoration proceeds from this custom. Whatever may have been pretended, it has nothing in common with the great monumental painting applied to the walls of temples or porticoes in the first period of Greek art. In order to convince oneself of this, it

---

[1] See a passage of Aristotle, quoted by Cicero (*De nat. Deorum*, II. 37).

is enough to study the manner in which the mythological or other scenes which ornament the Campanian houses are arranged upon the walls. They generally cover only a part of them, are placed amidst architectural decoration destined to set them off, distributed in regular compartments, and very often surrounded by a frame which seems to lean upon the wainscoat or rest upon consoles. The artist has evidently tried to produce a kind of ocular deception, and give to those looking at them the impression that these paintings are real scenes. The system of decoration can only be explained if we think of the habits and tastes of the Alexandrian epoch of which we have just spoken. We have seen that it had become a sort of mania among great personages to hang precious pictures on the walls of their houses. But the luxury is an expensive one, and not everybody could indulge in such costly fancies. It was necessary to be king of Egypt or of Syria, or at least a powerful minister or dreaded general, to have long oppressed the nations or unscrupulously pillaged neighbouring countries, in order to have these immense halls built, which historians describe with admiration, upheld by a hundred pilasters or a hundred marble columns, with marvellous statues before them, and pictures of the masters in the intervals. Citizens managed more cheaply. They had false pilasters painted in fresco[1] on their walls, framing false paintings, and in their small houses, when looking at the

---

[1] I have used the word "fresco" here and elsewhere, although certain *savants* consider it quite improper. Letronne denied absolutely that the ancient pictures were real frescoes, in the sense in which we now understand the word. M. Otto Donner, on the contrary, in a study

walls of their peristyle doubtless felt a pleasure like that of the kings and the great lords, when they walked in their palaces in the midst of masterpieces. Fresco was therefore an economical means, at the disposal of small folk, for imitating the example of the rich. As it requires rapidity of execution and allows of imperfections of detail, artists took advantage of it in order to work more quickly. They could work cheaper, and art became an industry. Petronius says: "That it was the audacity of the Egyptians which first invented the abridgment of the great art" (*Ægyptiorum audacia tam magnæ artis compendiariam invenit*),[1] and this opinion is very probable. It is natural that the country where people had the spectacle of the irritating luxury of great personages incessantly before their eyes, should also have been that where they sought more cheaply to procure themselves some of their enjoyments. Petronius adds that the employment of this cheap process ruined painting. This is also easy to understand. The poor, or, if you will, the less well-off, sought by some means or other to imitate the example of the rich; the rich, in their turn, were not long in borrowing from the poor. As the fresco painters usually attained from habit a sufficiently satisfactory execution, they ended by contenting themselves with the copies of famous pictures, and original painting was

---

preceding Professor Helbig's *Wandgemälde*, proves that the greater portion of the pictures that decorate the towns of Campania are painted in fresco. Without joining in the debate, which is not within my province, I consider myself authorised by M. Donner's conclusions to designate the Pompeian paintings by the name of frescoes.

[1] Petronius, *Sat.* 2. Professor Helbig has been the first to explain this phrase of Petronius satisfactorily.

no longer encouraged: hence the anger of critics and connoisseurs. Professor Helbig remarks that Pliny and Petronius express themselves on the subject of "this Egyptian invention" in the same tone in which certain amateurs of our day speak of photography, which they accuse of having ruined true art.

Everything, however, confirms the origin which Professor Helbig attributes to the paintings of Herculaneum and Pompeii. The pictures of which they are copies must indeed have belonged to the time of the successors of Alexander. They plainly bear the stamp of that epoch, and have all its characteristics. One of the great changes which then took place in the Greek world was that Monarchy almost everywhere replaced the Republic. Around the monarch and his wife, officers, ministers, servants, poets, and artists gathered—in short, a court was formed, and, as always happens, the influence of the court soon made itself felt in public manners. They became more polished, elegant, and refined. Distinction of manner, charm of mind, conversational refinement, and the delicate pleasures of society, were prized above all things. In great worldly assemblages where the two sexes meet, love naturally becomes the main interest, and so it assumed great importance in the society and consequently in the literature of that time. Poetry is henceforth to live upon it, and poetry will be imitated by the arts. But love, as usual, painted by the Alexandrian artists is not that mad passion which Euripides has represented it in his *Phædra*. Professor Helbig is right in saying that their painting no longer draws its inspiration from the Epics, like that of Polygnotus, or even from the ancient tragic theatre; it rather borrows its subjects from the idyll and the elegy, the favourite forms of Hellenistic poetry. With them, love is a mixture of gallantry and sentimentalism. They delight to represent goddesses and heroines afflicted by some amorous fortune. Œnone deserted by Paris, Ariadne on the shores of Naxos, gazing after the ship which bears away her lover, or Venus watching the hunter Adonis die in her arms. But they are careful that the grief of these beautiful deserted ones shall not impair their loveliness. Their despair has assumed very elegant attitudes; they are disconsolate, but adorned; they wear necklaces and double bracelets, and their tresses are fastened with golden nets. Besides, there is rarely wanting, in a corner of the picture, some little Cupid, to lend a more genial air to the scene when it threatens to become too severe. In the frescoes of Pompeii, Cupids are still more numerous than in the pictures of Watteau, Boucher, and other painters of our eighteenth century. They form the usual train of Venus, help her to adorn herself, hand her her gems, and hold the mirror in which she gazes. They bring her to Mars, who awaits her; they surround wounded Adonis, draw aside his garments, and bear his crook and his spear. It is again a Cupid who leads Diana into the cave of Endymion and shows her the beautiful youth asleep. When Œnone seeks by her despair to detain her faithless spouse who is about to leave her, Paris is indifferent to her reproaches, and seems scarcely to listen to them: and with good reason, indeed, for the artist has represented behind him a Cupid bending caressingly towards his ear, and doubtless discoursing to him of his new passion. In these different paintings the Cupids are only accessories, but there are others where they form the entire picture. We are shown them all alone, and engaged in occupations which are usually the lot of man. They dance, they sing, they play, they feast; with whip upraised, they drive a chariot drawn by swans, or endeavour with great pains to guide a team of lions. They gather grapes and grind corn in a mill, aided by little donkeys, which they lead with garlands of flowers. They sell, they buy, they hunt, they angle; and this last amusement appears to our artist so divine a recreation, that they several times attribute it to Venus herself. One of the most pleasing pictures, and the best known, in this fanciful and coquettish style is the Cupid-seller. An old woman has just caught a little Cupid in a cage, and, holding him up by his wings, offers him to a young girl, who desires to buy him. The latter does not seem to be entirely a beginner, for she already holds another Cupid upon her knees. She, nevertheless, looks with great curiosity at the one she is going to buy, and who stretches his hands joyously towards his new mistress.

I have already said a word about what became of mythology in the new school of painting. We saw that the old myths lost their deep and serious meaning. One of the usual processes of these painters, when they take up subjects to which ancient art had lent an ideal size, is to reduce them, as far as they can, to human proportions. They like to entirely efface the distance separating gods from men, and to treat heroic legends like adventures of everyday life. It is evident that, in painting the loves of the gods, the artist always had before his eyes what passed at the courts of the Seleucides or of the Ptolemies. In the famous "Judgment," Venus, who wishes to be preferred, coquets with Paris like a woman of the world; while Polyphemus, seated on the sea-shore, sings his sorrow to his lyre, a Cupid is seen arriving on a dolphin, with a letter to him from Galatea. Mars and Venus are prudent lovers who would fain not be discovered while engaged in their sweet encounters. They are shown by a Pompeian painter cautiously guarded by a dog, in order that they may be warned of the approach of the indiscreet. This is indeed a very vulgar way of introducing real life into heroic legends.

The character of these paintings clearly shows their age; it is truly Alexandrian art which we have before our eyes, but is it certain that this art was faithfully reproduced in the Pompeian frescoes, and how far may they be used as a means of judging it? This is a delicate question which Professor Helbig has handled in a very interesting manner. The first shows, by a study of the very conditions of painting at Pompeii, that there must be inevitable differences between the originals and the copies. The Pompeian houses are usually small; the space which the architect gave over to the painter was not, as a rule, of great extent, and was little suited for what the Greeks styled "megalography." In the arts, dimension has much importance, and often great subjects, when confined within too small a frame, become character pictures. This is what happens at Pompeii, where the frescoes are generally nothing more than reproductions of larger and more extensive compositions. Let us add that if these frescoes appear to us somewhat lacking in variety, the fault was not altogether imputable to the Alexandrian school, whence they proceed. Among the innumerable subjects offered to them by this school, the Pompeian artists were forced to make a choice. They took, in preference, cheerful and gay scenes, avoiding those which appeared to them too mournful. "A violent picture overwhelms the soul," said Seneca.[1] These good townsfolk who desired to live joyously in this happy country, at the foot of the verdant slopes of Vesuvius, would not have liked all the horrors of ancient mythology to be presented to their gaze. The crimes of the family of Agamemnon, the death of Hippolytus, torn by the thorns of the way, had, as we know, given rise to famous pictures of the Alexandrian artists. We do not find them again at Pompeii. They were not in their place in the halls reserved for calm family joys. When Pompeian artists ventured to paint some less pleasing scene, they, for the most part, modified it. Circe bound to a furious bull, and Actæon devoured by his dogs, are with them only pretexts for studies of nude women and agreeable landscapes. In reproducing a picture in a fresco, it is inevitably perverted. Fresco does not in the same degree allow of that refinement of touch and perfection of detail which were the chief qualities of the Alexandrian masters. These, too, were not the qualities mainly sought after by the Pompeian painters, and it may even be maintained that they did not need them. Now that the houses of Pompeii have no longer roofs, we see them

---

[1] Seneca, *De ira.*, II. 2.

illumined by a dazzling sun which brings out their least defects, but they were not made for this glare. The halls where they are placed were usually only lighted by the door, and precautions were even taken that all the glow which flooded the atrium should not enter by this one opening. Awnings stretched from one column to the other made a shade before these chambers, where the inhabitants passed the hot hours of the day. In this half-darkness, imperfection of detail did not appear, and the artists could, without impropriety, neglect some of the merits of the models they copied.

In spite of these reserves, which are indispensable, it may be admitted without rashness that the frescoes of Herculaneum and Pompeii give a sufficiently correct idea of Alexandrian painting. Professor Helbig is so convinced of this that he would fain identify again, in these incomplete copies, some of those renowned paintings whose beauty has been extolled by ancient critics. This is an undertaking which may at first sight appear somewhat hazardous, but it must not be forgotten that if these pictures are now lost, we at least retain some mementoes of them. They are mentioned in the writers who have handed us down the history of ancient painting; the poets, and especially those of the *Anthology*, have rarely omitted to devote a few lines to a description of them; more or less exact copies of them are found in bas-reliefs, and on vases ; and, in fine, what is most important, they must have been often reproduced on the walls of the Campanian towns. On comparing these different copies, and controlling them by the information afforded us by critics and poets, we perceive what each artist has taken from the originals, and we get to reconstruct it, at least so far as regards its entirety and its great lines. Thus it is that, by an effort of science and sagacity, Professor Helbig gives us back two famous pictures of Nicias, his "Andromeda" and his "Io." The first is twice reproduced at Pompeii, in proportions that are not usual there; the other only appears once, but has fortunately been found again in the house of Livia, on the Palatine. They are two fine pictures which appear made to match each other, and are sufficiently alike for us to think them by the same hand. The copyists must have adhered to the general arrangement and chief qualities of the original, and they therefore allow us to picture to ourselves what those two works of the Athenian artist must have been, who, according to Pliny, excelled in painting women. This is also what happens in connection with a picture still more famous than those of Nicias. Two small frescoes at Pompeii represent Medea about to slay her children. Scholars agree in admitting that they are imitations of a masterpiece of Timomachus, but rather imperfect imitations. Beside Medea, the painters have placed her two sons, playing a dice, under the supervision of their tutor. This dramatic detail, this striking contrast between the careless joy of the children and the terrible preoccupation of the mother, evidently belong to the original picture. The remainder, in the Pompeian frescoes, is less happy, Medea's face, especially, being wanting in character. Fortunately, a "Medea" was found at Herculaneum of larger dimensions, and which displays a more assured talent. This time she is represented alone, without her children, with mouth half open and eyes distraught.[1] Her fingers clutch the hilt of the sword with a convulsive movement, she appears to be torn by unspeakable sorrow. This figure, one of the finest that remain to us from antiquity, is certainly by a painter of genius. It is a thing the Pompeian copyists would not have imagined; the hand of a master is found in it. Thus, by placing the group of children from the Pompeian frescoes beside the Medea of Herculaneum, we are sure to have all the picture of Timomachus.[2] It is therefore the whole of an important epoch of Greek art which has been preserved for us in this corner of Italy. The pleasure we feel in looking on these pictures increases when we reflect that they alone represent a great school of painting, but this certainly does not mean that they only interest us because they recall lost masterpieces, and that they are unworthy of being studied for themselves. I fear, lest by dint of repeating the words "imitators" and "copyists," we may have unduly depreciated the merit of these unknown artists. To be content to call them decorators is not to them justice. They doubtless imitated, yet with certain independence. They were not wholly the slaves of their models, but interpreted them freely, modifying them according to the conditions of the places they were to paint, or the humour of the master they had to please. What proves this beyond doubt is that a great number of replicas

---

[1] Ovid (*Trist.* II. 526) seems to recall the picture of which we speak when he says : *Inque oculos facinus barbara mater habet.*

[2] We have the proof that the Medea of Herculaneum, which decorated a very narrow space of wall, was detached from a larger fresco. The picture of which it originally formed part very probably contained the children and their tutor.

are found at Pompeii, evidently from the same originals but never alike. So something personal entered into the work of these artists, which fostered their talents, prevented them from being mere artisans, and made them veritable painters. It was this which enabled them to invent for themselves when needful. They rarely did so, being forced to work quickly, and finding it more expeditious to borrow from others than to take the pains to create. We have remarked, however, that they sometimes took their inspirations from scenes of which they had been witnesses, and painted character pictures of inimitable truthfulness. But, whether imitating or inventing, they do all with an ease and grace, rapidity of execution and sureness of hand, which we cannot help admiring. Our admiration increases when we remember that they worked for the burghers of a small town, and, above all, when we reflect that in all the Roman world there must have been the same tastes as at Pompeii, and that there must have been artists everywhere capable of the same work. This is what astonishes and confuses our minds. Historians tell us that there were no longer painters of genius at that time, but the paintings of Pompeii show us that painters of talent were never more numerous. In our days, we boast that we put ease within the reach of the greatest number, and popularise well-being. This is a great boon. In the first century, something of the kind had been done for the arts. Thanks to these convenient processes, which permitted masterpieces to be disseminated, they had ceased to be the privilege of the few, in order to become the pleasure of all.

IV.

WHENCE THE RESEMBLANCES COME THAT ARE REMARKED BETWEEN THE PAINTINGS AT POMPEII AND THE POETRY OF THE AUGUSTAN AGE—THE PAINTERS AND THE POETS INSPIRED BY THE SAME SUBJECTS—LATIN LITERATURE IMITATES THE POETIC SCHOOL OF ALEXANDRIA — CATULLUS —VIRGIL — PROPERTIUS—OVID—DIFFERENCES BETWEEN THE PAINTERS OF POMPEII AND THE ROMAN POETS—THE PAINTING NEVER BECAME ROMAN — REPUGNANCE OF THE POMPEIIAN ARTISTS TO HANDLE SUBJECTS DRAWN FROM THE HISTORY OR THE LEGENDS OF ROME—IS POMPEII REALLY A GREEK TOWN?—NATIONAL CHARACTER OF THE POETRY OF THE AUGUSTAN AGE.

ON closely studying the Pompeian paintings, we are forcibly struck by the resemblance they bear to certain poems of the great epoch of Latin letters, and, above all, to those of the elegiacs or didactics who sing of mythology or of love. Among the poets, as among the painters, the same subjects are incessantly reproduced, and they are treated in nearly the same manner. They both love to express the same feelings, affect the same qualities, and omit to avoid the same effects. Must we conclude from this that the painters were inspired by the poets, and took the subjects of their pictures from their works? We shall presently see that this was not so, and that it is easy to demonstrate that they remained almost entirely strangers to the literature of Rome. Must we believe, on the contrary, that it was the poets who imitated the painters? This supposition would scarcely be more probable, and, at any rate, it is needless. We have a very simple means of explaining all. If they resemble each other, it is because they drew from the same source. Painters and poets worked both from the same models. They were pupils of the masters of Alexandria, and this is how they might happen to meet, without even being acquainted with each other.

We know that the Romans do not possess a truly original literature, and that they always lived by borrowing. They began by imitating the classic poetry of Greece, *i.e.* that which flourished from Homer down to the time of Alexander. This, it must be owned, was to choose their models well, yet I do not think they must be allowed too much credit for their preference. In those remote times they were scarcely in a condition to distinguish ancient from modern Greek literature, the writers of the Periclean age from those who flourished at the court of the Ptolemies. The choice they then made is explained less by the refinement of their taste than by existent circumstances. The old Greek poets, although in society somewhat thrown into the shade by the glory of new writers, continued to reign in the schools with undivided sway. The grammarians explained them to their pupils and made them the foundation of public education. As the Romans first knew Greece through the intermediary of the professors who came to educate their children, they were naturally led to admire and imitate the writers who were admired in the schools, that is to say, those of the classic age. It must also be said that these old poets, by their grandeur and their simplicity, suited an energetic and youthful race about to conquer the world. Unhappily, the virile virtues of the first Romans were not proof against their fortunes, and at the moment when the former began to wane, the very progress of their conquests brought them into a more direct contact with the Greeks. Having become acquainted with Greece in the schools and by means of books, they were about to see it at home and to frequent it habitually. At Athens, at Pergamus, and at Alexandria, those great towns they so delighted to visit, and of which several had been the capitals of powerful kingdoms, they found an enlightened, polished, and witty society, in whose midst they were happy to live with a literature differing from that taught them by their masters, and which charmed them at once. In vain friends of the past resisted. Cicero complained bitterly of "these lovers of Euphorion" who dared to banter Ennius and preferred a wit of Alexandria to him. Lucretius also remained faithful to Ennius and the ancient poets, acknowledged them for his masters, and loved to imitate their vigorous and sober verses; but the new school had on its side what brings success, youth and women. Those beautiful freedwomen, who reigned in fashionable assemblies and governed political men, loved to repeat the verses of Calvus and Catullus. Thenceforth the imitation of the Alexandrians slips in among nearly all the poets, and it especially prevails in Ovid and in Propertius, who proclaims himself, without ambiguity, the pupil of Callimachus and of Philetas.

This is why the Roman elegiacs have so often come

face to face with the painters of Pompeii. These resemblances are not simply curiosities, agreeable to remark in passing. Professor Helbig thinks that a serious interest attaches to them, and that they may help us to a knowledge of the literature of the Augustan age. The poets of Alexandria being lost, it is difficult to say how far those of Rome faithfully produced them, and to distinguish what they borrowed from what is really their own. In order to ascertain this, let us compare them with the painters of Pompeii. When their descriptions faithfully recall some Pompeian picture, we shall conclude that the painter and the poet had a common model before them, and that they are both imitators.

We do not know to whom Catullus owes the most beautiful of his poems, that in which he represents Ariadne deserted by Theseus and consoled by Bacchus. M. Riese thinks that he translated it from Callimachus,[1] but he has given no positive proof of this. What is certain is, that this subject is often found reproduced on the walls of Herculaneum and Pompeii, and that, consequently, it must have been very common among the poets of Alexandria. Furthermore, Catullus has certainly treated it in the Alexandrian manner. Mixing with strokes of deep passion many graceful diminutives, he does not, in this terrible moment, neglect to describe his heroine's toilette; to speak to us, by the way, about her fair hair and her charming little eyes; and lastly, to tell us that when she advances into the waves to try and follow her

---

[1] *Rhein Museum*, XXII. p. 498.

fleeing lover, she is careful to raise her dress to her knees—

"*Mollia nudatæ tollentem tegmina suræ.*"

Virgil also began by yielding to the taste of the moment and imitating the Alexandrians. This is what explains the defects imputed to his first works. In his *Bucolics* some blemishes are found surprising in a mind so exact and subtle. The Arcadian shepherds who inhabit the shores of the Mincio, the statesmen become graziers, who weave reed baskets in solitary caves, and play upon a rustic pipe, to console themselves for the infidelities of an actress who has followed an officer, this manner of transferring to the country the events of the town, and of putting political allusions among pastoral discussions, reminds Doctor Helbig of the strange fancies of certain Pompeian landscapes, where town and country are found oddly mingled together,—elegant porticoes in the solitude where Polyphemus leads his flock to pasture, and an Ionian temple, crowned with garlands on the heights of Caucasus, near the vulture who devours Prometheus. In Propertius, the influence of the Alexandrians is yet more visible still, and his elegies show more points of resemblance with the paintings of Pompeii than do Virgil's Eclogues. Of mythology it is brimful. Whether sad or joyous, all his sentiments are expressed by allusions to the ancient legends. He has no more delicate eulogy for his mistress than to compare her with the heroines of ancient times. If he one day surprises her asleep with her head leant upon her arm, she immediately reminds him of Ariadne stretched upon the shore of Naxos,

Andromeda after her wondrous deliverance, or the exhausted Bacchante who falls overcome by invincible sleep on the plains of Thessaly. These are personages well known to those who have visited the Campanian towns, for they are found there everywhere. When Cynthia, after a long resistance which has much distressed the poet, at length yields to his love, he celebrates his victory by an explosion of mythology: " No, the son of Atræus was not more joyful when he saw the fortress of Troy fall at his feet. Ulysses, after all his journeying, did not land with so much pleasure on the shores of his cherished isle. Electra, when she beheld her brother, whose ashes she thought she held in her hands, or the daughter of Minos, on seeing Theseus, whom she had just delivered from the labyrinth, did not feel as much happiness as I knew last night. Let her but once again grant me her favours, and I shall hold myself immortal!"[1] The little Cupids we have so often found in the Pompeian paintings are not wanting in the poetry of Propertius either. When he decrees himself a sort of triumph for having acquainted the Romans with the Alexandrian elegy in all its beauty, he joins the Cupids in it, and wills that they shall take their places in the same car as himself.

" *Et mecum in curru parvi vectantur Amores.*"[2]

In one of his most pleasing pieces, imitated by André Chénier, he relates that one night, after a debauch, he wandered alone with unsteady steps about the sleeping own in search of some guilty love adventure. Suddenly, he falls into the midst of a troop of little children

---

[1] Propertius, II. 14.   [2] *Ib.*, III. i. 11.

whom his fright prevents him from counting. "Some carried tiny torches, others held arrows, and others again seemed to be preparing bonds to bind me. All were naked. Anon, one of them, more resolute, cried: 'There he is! seize him; you know him well; him it is that an angry woman has charged us to give her back." He spoke, and already I felt a knot press my neck; the rest approach: 'Chain him, scold him, and take him back repentant and happy to Cynthia's house."[1] Is this not the subject for a charming picture which might be placed opposite the "Cupid-seller"?

But it is Ovid especially who seems most to have profited by the poets of Alexandria, so it is also he who most frequently recalls the Pompeian paintings. It would be easy to choose among these pictures a certain number which might serve as it were for illustrations to his works, so much the poet and the painter at times resemble each other. They represent Io delivered by Mercury, Hercules spinning with Omphale, Paris carving Œnone's name on the trunks of trees, Europa "holding the bull's horn with one hand and resting the other on his back, while the wind shakes and swells her garments." Further back I have mentioned the picture where the disconsolate Polyphemus receives a letter from Galatea, brought to him by a Cupid mounted upon a dolphin. This strange invention at once sets one thinking of Ovid's *Heroïdes*. These are love-letters which lead us to suppose not only that people knew how to write, and wrote much, at the time of the Trojan war, but that there were then means of forwarding epistles

---

[1] Propertius, II. 29.

even when addressed to persons whose whereabouts were unknown, or when one was consigned to some desert isle. These ways are little in keeping with such far-off times. For it to be conceivable that women should write such long letters, fraught with such brilliant thoughts and so much knowledge of the human heart, it must be supposed that great pains were taken to bring them up well. So the poet says in express terms that they had had masters, and that they had been taught the arts which are the ornament of infancy.[1] "In reality, they are only contemporaries of Corinna," who had frequented good society and learned the usages of gallantry in "The Art of Loving." It is Ovid's usual system as much as possible to rejuvenate this ancient mythology, and the gods do not escape any more than the heroes. With him they quite lose that antique air which rendered them venerable. He makes men of them, and men like those among whom he passed his life. Hercules is no longer anything more than a common athlete who fights Acheloüs after the same fashion as those shown to the people in the public games.[2] When Minerva defies Arachne, she sets to work like a buxom workwoman, tucking up her gown in order to be less encumbered, and setting her shuttle flying among the threads with an ardour that makes her forget her grief.[3] Jupiter's household is entirely wanting in gravity, and Juno is continually busied in looking after her husband, who gives her great cause for jealousy. This custom of representing the gods quite like men, and giving antique mythology

---

[1] Ovid, *Met.*, IX. 717. [2] *Ib.*, IX. 36. [3] *Ib.*, VI. 60.

a modern air in order to render it lifelike, we have also remarked in the paintings of Pompeii. It is a proof that it already existed among the poets of Alexandria. But Ovid goes much further than his masters. He mingles with everything a species of humour and buffoonery not at all in keeping with the spirit of the Alexandrians. While imitating, he has deeply changed them. M. Rohde, in his book on Greek romance, remarks that if he owes them the foundation of his works, he differs from them in the execution.[1] The Alexandrians were for the most part scrupulous and exact, critics as much as poets, very severe towards others and towards themselves, who, desiring to please people of the world, expended great care on their verses, polished and chiselled their phrases, strove to put in wit or knowledge everywhere, and consequently produced but little. They certainly had a pupil in that Helvius Cinna, the friend of Catullus, who took nine years to finish a poem, and by dint of working at it, rendered it so obscure that he immediately had commentators, and to understand it came to be considered a glory. Ovid was not one of those syllable-scrapers, one of those fastidious ones who are never content with themselves. He was quick of thought and swift of hand, and his pleasure and his talent were to improvise. He charmed the society in which he moved, not only by following its tastes and flattering its caprices, but by continually dazzling it with new works. It may also be said of him that he supplies the place of those " room pictures " of the Alexandrian school—so careful,

---

[1] E. Rohde, *Der Griechische Roman*, p. 125.

so finished by bold frescoes, full of unpleasing negligences and defects—but in which we find a fertility of resource, a wealth of detail, and a rapidity of execution that charm even those most inclined to be critical. This is one resemblance more with the painters of Pompeii.

But these painters and poets do not always resemble each other. Some differences also exist, to be noted with care, since they complete our knowledge of them. I do not mean only those arising from the varied conditions of their arts; these could not be escaped, and recur everywhere. When Horace says that poetry is like painting, he does not mean to express an absolute truth subject to no exception. That acute critic well knew that if their aim is identical, they follow different ways in order to reach it. Painting, which works directly for the eyes, is forced indeed to give its personages fine attitudes. It can present nothing to the view that would shock it, for the image does not become effaced; the impression would remain and grow more irksome from its very duration. The poet, on the contrary, who appeals to the imagination, and paints with a stroke, may allow himself vagaries which would not be allowed in a painter. I will only instance a single example. The legend relates that Io changed into a cow, and it is under this form that she is pursued by the anger of Juno, who confides her to the watchful guard of Argus, the shepherd with the hundred eyes. Ovid accepts the legend as it is, changing nothing and concealing nothing. On the contrary, it amuses and pleases him, and its oddity is just what he develops with greatest zest. He depicts the unhappy Io still unconscious of her metamorphosis. "She would fain implore her keeper and stretch forth her hands to him," but no longer finds an arm to stretch.[1] She strives to speak, but her words are bellowings which frighten her. She approaches a fountain where, in happier times, she was wont to mirror herself, but directly she sees her horns she flies, terror-stricken, before her image. All this is said delicately, with a tone of pleasing irony, without taking into account that Io's father himself, in spite of his grief, cannot refrain from a comic reflection. "And I," says he, "who sought thee a spouse, and thought to give myself a son-in-law and grand-children, it is in my herd that I must choose thee a husband, it is in my herd that I must look for grand-children." A painter could not permit himself these pleasantries. It would be difficult for him to excite our compassion for a cow, to interest us in its misfortune, and make us desire its well-being. Therefore, in spite of Juno, Io will remain for him a beautiful young captive girl, watched over by a wicked gaoler, who raises her eyes and stretches her arms towards heaven to invoke a deliverer. At the very utmost, the most scrupulous artists, determined at any cost to respect the tradition, will draw upon her charming brow two little horns, half concealed by her tresses, and this is the only memento that the metamorphosis of the daughter of Inachus will leave in a picture. It is the same with respect to her keeper. The hundred eyes bestowed on him by the legend

---

[1] "*Illa etiam supplex Argo quum brachia vellet,*
Tendere, non habuit quæ brachia tenderet Argo."
—Ovid, *Met.*, I. 629.

greatly exhilarate Ovid, who congratulates him on being able to turn about as he lists without losing sight of his victim—

"*Ante oculos Io, quamvis aversus, habebat.*"

Let us suppose the painter resolved to adhere faithful to the legend, he will never produce anything but a grotesque figure. He escapes the difficulty by making Argus like an ordinary shepherd, and contenting himself with putting upon his shoulder a leopard skin, whose spots must represent in the eyes of a complacent spectator the hundred eyes of Argus. This is how the painter grapples with difficulties non-existent for the poet, but which sometimes oblige the former to treat the same subject in a different manner.

These difficulties, I repeat, were inevitable, since they are inherent in the very conditions of the two arts, which cannot be changed, so it is needless further to insist upon them. But there is another more important, and which deeply separates the painters of Pompeii from the Latin poets. The other arts which Greece gave to Rome seem to have made some effort to acclimatise themselves in their new country, and to have in some degree adopted its qualities and its character, but painting never became Roman. Not that it had more cause than the other arts to complain of the welcome received from the Romans. From the day when Paulus Œmilius sent for Metrodorus from Athens to paint the pictures that were to adorn his triumph, and charged him to educate his children, great artists found consideration and fortune in Rome. Fine pictures were paid for at as high a price as the statues of the masters, and if they were very anxious to fill the public places and the porticoes with marble or brazen images of the gods or of great men, they were not less so to adorn public or private monuments with frescoes, and the example of Pompeii shows how general this taste became. What proves that painting was not without honour in Rome, even in the most ancient times, is that it was one of the first arts practised by the Romans themselves. Before the time of the Punic Wars, a patrician, a man belonging to an ancient and illustrious race, did not disdain to become a pupil of the Greek artists, and decorate a temple with his own hand. His talent made him so renowned that thenceforth he was only called Fabius the Painter (*Fabius Pictor*), and his family kept the name. From that moment downward, Romans are not wanting in the lists of painters who made themselves famous, and among those whose memory Pliny has preserved for us, there is one who was so proud of his country that he never quitted the toga, even when he had to climb some scaffolding,[1] much in the same way that Buffon is said to have donned court dress while composing his great work. But whether he wore the toga or the pallium, the artist remained Greek. Greek painting did not change its method on settling in Italy, modified its habits in nothing, and sought its inspirations only in the memories of its former country. Letronne is right in saying "that it was a plant which flourished everywhere as on its native ground, scarcely feeling the influence of the change of soil and climate."

---

[1] Pliny, XXXV. 37.

It is indeed thus that it appears to us at Pompeii. It is surprising to see to what a degree painters who worked in an Italian town, for people who delighted in nothing more than to be called Roman citizens, and at a time when the sentiment of national glory was more alive than ever, remained unaffected by the influence of Rome. While, by their side, sculptors, also Greek in origin, delighted to people the public places with images of the imperial family, they never thought of painting in the temples decorated by them the exploits of Augustus or of his successors. The history of Rome—that glorious history which astonished all the world—never inspired them. In their mythological paintings, the subjects are always borrowed from Greek traditions and legends. Above all, there was at this time a great Roman poem, consecrated by public admiration, which all the world knew by heart, at Pompeii as much as elsewhere, for we have proofs of it—Virgil's *Æneid*. This work, in so many ways connected with the Homeric Epos, was not of a kind to displease Greek artists. They did not find themselves abroad in a poem in which Greece is everywhere present, and whose heroes are borrowed from the *Iliad*. Yet among all the paintings of Pompeii, only five or six pictures have been found whose subjects were taken from the *Æneid*, and even of these one is a caricature. It represents a young long-tailed monkey, covered with a coat of mail, encumbered with a sword, carrying an old monkey on his shoulders, and dragging a young one along by the hand: it is Æneas, leaving Troy with his father and his child. Of all the rest, a single one has some importance. This is a very faithful imitation of a scene from the twelfth book of the *Æneid*. Æneas, struck in the course of the battle by an arrow, leans one hand upon his javelin and the other on the shoulder of his weeping son, yielding his leg to the physician, old Iapyx, who endeavours to draw the dart from the wound. Above him, Venus, his mother, descending from the sky, brings the *dittany* that is to cure him. It is not a good picture. The attitudes of the personages are awkward, the whole is wanting in ease, and it is seen that the subject not being familiar to the artist, he did not treat it with pleasure. The adventure of Dido, which appears made in order to tempt a painter of talent, has only been two or three times represented at Pompeii. This is little, indeed, it must be owned, especially if we reflect that the story of Ariadne deserted by Theseus, which so resembles that of Dido, has given rise to more than thirty works, some of them of large dimensions and remarkable execution.

It has, it is true, sometimes been said that Pompeii was more a Greek than a Roman town, and that in excavating for its works inspired by the legions and traditions of Greece, the artists followed its taste. But this opinion, although it has been widely spread,[1] is none the less very incorrect. From the time when they received the right of citizenship, the inhabitants of Pompeii considered themselves Romans. Latin is not only their official language, used by magistrates

---

[1] Mazois said at the beginning of his second volume: "It will perhaps cause surprise that I class the houses of Pompeii with Roman dwellings, for this kind of Greek taste which prevails in the ruins seems to have accustomed everybody to look upon the houses of the town as Greek."

in decrees—it is the common idiom, that of the poor as well as of the rich; not only of peasants but of townspeople, used no less in private than in public life. Children who chalk their jokes on the walls, young folk addressing a salutation to their mistresses, idlers who celebrate their favourite gladiator on leaving the public games, and frequenters of taverns or of questionable resorts who want to describe their impressions, do it in Latin almost exclusively, Oscan and Greek being always the exception. The Pompeians not only speak the language of their masters, but share all their feelings. The Emperor has no more devoted subjects, and they were the first to adopt the worship of Augustus. Doubtless we are not surprised that official inscriptions should be full of expressions of respect and affection for the sovereign and his family. What astonishes us more is to see that those chalked upon the walls by men of the people, who can be less suspected of flattery and untruth, often contain similar protestations. Thus we often find the cry of "Life to the Emperor" (*Augusto feliciter*), and one of those who writes upon the wall adds this thought: "That the health of princes makes the happiness of their subjects" (*Vobis salvis felices sumus pertetuo*).[1] Another must needs send to Rome, the former enemy, a distant salutation: "*Roma vale.*"[2] If the masterpieces of Greek literature are not unknown at Pompeii, Roman literature is still more current there. Cicero is sufficiently read for him to be parodied,[3] while

---

[1] *Corp. insc. lat.*, IV. No. 1174.   [2] *Ib.*, No. 1745.

[3] It is impossible not to recognise in a very light inscription (*Corp. insc. lat.*, IV. No. 1261) a parody on a famous passage of the *Verrines*.

Propertius, Ovid, and even Lucretius, are quoted continually. But it is especially the *Æneid* that appears to have been the pleasure and the study of all. Virgil had interested the whole of Italy in his work by singing all its memories and all its glories. From Pompeii that point of Misenum was visible, the tomb of one of the companions of Æneas, which the poet had mentioned in his works, and it was near those Phlegrean fields where he had placed the entrance to Hell. Thus the knowledge of the *Æneid* was very widely spread among the Pompeians of all classes. What well shows this is that inscriptions scribbled upon the walls, which can only be the work of schoolboys or men of the people, often contain verses of it. It was known by heart, people loved to quote it, and even the unlettered had some acquaintance with it. But it is probable that in a town where Virgil seems to have been so popular, people would have liked to have some of the scenes described by him represented upon the walls of their houses. If the painters have hardly ever done so, if they have rarely placed before the Pompeians subjects borrowed from their favourite poet, or mementoes of their natural history, it is because the art which they practised had remained Greek, that it was known to be shut up in its traditions and its habits, and nobody required it to leave them.

With poetry it was not the same, and it is this which distinguishes it from painting. Although it also came from Greece, it consented with a good grace, and almost at once, to become Roman. Nævius uses the forms of the Homeric Epos for the purpose of celebrating the heroes of ancient Rome, while the tragic method of Sophocles lends itself to sing the exploits of Decius, of Paulus Æmilius, and of Brutus. This mixture reached its perfection in Virgil's *Æneid*. Nowhere did the traditions of the two countries, the genius of the two peoples, and their two antiquities, more harmoniously blend them in Virgil's poem, and this it is which makes its admirable beauty. At that moment Rome appears prouder than ever of her past, and more busied with her history. The Emperor, who has taken away its freedom, stimulates its national pride. In order to fill its imagination and assuage its regrets, he constantly shows it the immensity of its territory, extending to the limits of the civilized world, and recalls the heroic manner of its conquest. In order to dissimulate the newness of its institutions he surrounds himself with the great men of ancient times, puts himself in their company, and boldly presents himself as their follower. A kind of watchword then went forth to all the poets to unite in the eulogy of the prince that of the heroes of the Republic and the memories of ancient Rome. Not one of them neglected to do so. Even the most trivial, who had never busied themselves with aught save their amours, assumed a grave tone, and mingled their light verses with patriotic songs. Propertius, a prudent man, had settled the employment of all his life in advance. He purposed: "When age should have chased away pleasures and sown his head with white hairs, to enquire out the laws of nature, find how this great house of the world is governed, study the principles which direct the course of the moon, whence come eclipses and storms, why the rainbow drinks the waters of the rain, and what is the cause of the underground agitations which make the high mountains tremble."[1] In other terms, he meant to remain a true Alexandrian to the end of his days, only proposing to pass with age from the elegies of Callimachus to the didactic poetry of Aratus. Yet he did not withstand the solicitations of Mæcenas, and he, too, ended by celebrating the ancient traditions of Rome, "and putting all the breath that issued from his feeble bosom at the service of his country." Thus it is that the elegy—that is to say, the species of poetry which the Romans had most directly borrowed from the Alexandrians—at length mixed novelties with its imitations, and by dint of celebrating the great memories of the national history, became Roman in its turn.

So it may be said, with perfect truth, that the Latin poetry of the great age is best appreciated by comparison with contemporary painting. They issued from the same source, but they took different ways, and they elucidate each other both by their relations and their differences. When we see with what obstinacy painting remained entirely Greek, we render better justice to the efforts made by poetry to adapt itself to the country into which it had come to settle. These efforts gave it an element of strength and life which it is not possible to mistake. By becoming Roman, by flattering the national pride, and by endeavouring to respond to popular feeling, it rendered its action upon the crowd more potent. In this respect it was original and owed nothing to the school of Alexandria, which

---

[1] Propertius, III. 5, 23.

never had these patriotic impulses. As for all that mythology which it too easily borrowed, and which we now find so flat and obscure, the Romans must assuredly have taken less interest in it than the Greeks, with whom it was born; but to think it was quite indifferent or unknown to them, were to mistake. Painting had long since popularised it among them. It is impossible to ascertain at what point of time Greek artists entered Rome and began to exercise their calling there, but it must have been early. Plautus tells us of pictures which in his time adorned private houses, and which represented Venus with Adonis, or the eagle carrying off Ganymede.[1] In Terence, a lover who hesitated to commit a very bad action, relates that he lost all his scruples after seeing on the walls of a temple Jupiter seducing Danæ.[2] These are the subjects most often found in the towns of Campania. Thus, for several centuries, painters had ornamented public and private buildings with them; the eye and the mind had become accustomed to see them; even the ignorant and unlettered were grown insensibly familiar with them, and poetry, which was to take them up in its turn, found a public already prepared in advance, and much more extensive than is believed. Then something happened much like what took place with us, when the tragic poets of the seventeenth century put Augustus or Agamemnon upon the stage. These Greek and Roman personages were not strangers to the spectators. Classical education, on which all

---

[1] Plautus, *Menæchmi.*, 1, 2, 34.　*Merc.*, 2, 2, 42.

[2] Terence, *Eun.*, 3, 5, 36.

France was formed, rendered these names familiar to those who frequented the theatre. The clerk who for fifteen sous purchased the right to kiss Corneille was as well acquainted with them as the magistrates and the great lords. People knew their history better than that of the heroes of ancient France, and were more familiar with them. Some critics imagine that by treating subjects taken from antiquity, our poets condemned themselves to work for a small number of persons. This is a great mistake: they addressed all. The schools had made them a vast public, prepared to understand and disposed to applaud.

V.

THE BURGHERS OF POMPEII — THE POOR — WHERE DID THEY LIVE? — INNS AND TAVERNS — OCCUPATIONS AND PLEASURES COMMON TO THE POOR AND THE RICH—THE MUNICIPAL ELECTIONS—THE SHOWS—HOW MAY WE BECOME ACQUAINTED WITH THE INNER LIFE OF THE POMPEIANS?—THE INSCRIPTIONS AND "GRAFFITI"—THE SERVICES THEY RENDER US.

FROM these general considerations, which have in some measure turned us aside from our subject, let us go back to Pompeii and its inhabitants. The paintings which I have just been studying at such length, and which teach us so many things concerning ancient art, also give us some curious information about the city where they were found. Although it may be said that painters were then extremely numerous, and worked very cheaply, it is still clear that in order to think of having one's apartments decorated with elegant frescoes, it was necessary to be in the enjoyment of a certain ease. On this reckoning, there must have been a great many well-to-do people at Pompeii. The considerable number of houses containing interesting paintings proves how fortune was spread there. Moreover, all the studies hitherto made lead to this conclusion. M. Nissen, the patient investigator of these ruins, has confirmed that from the time of the Empire the taste for luxury at Pompeii seems to grow year by year. Private houses become more and more beautiful and ornate, while public buildings are continually enlarged. What he called the monument fever (*Denkmalsfieber*), "one of the chronic maladies of ancient democracies," progressed there every day.[1] Parvenus desired to display their sudden fortunes by building or repairing temples, and they caused statues to be decreed to them by the councils of the town, or the corporations whom they protected. Beside the ancient nobility, there rose a citizen class, rich, important, jealous of consideration, partial to splendour and pomp, and, above all, very numerous, which lived freely, loved well-being, and was desirous to allow itself some of those enjoyments and privileges hitherto reserved for the great families.

And the poor? they doubtless existed at Pompeii, as elsewhere—more than elsewhere, perhaps. It was, as we have seen, an industrious town, and one which did a great deal of business. Independently of its maritime commerce, it produced in abundance wine and fruit, which it exported to the other towns of Italy. Pliny and Columella tell us that, especially, its cabbages were

---

[1] *Pompeianischen Studien.*, p. 373.

famous. A kind of sauce or seasoning called *garum* was also made there, with salted fish, which was the delight of epicures. It is natural that in a commercial town there should have been a great number of workmen. At Pompeii, as everywhere else, they had their regulations, their feasts, and their places of meeting. We know those of the goldsmiths, the wood merchants, and the muleteers, who take part in the elections and recommend their candidates.[1] It has also been conjectured that the cloth factories, fulleries, and dyeing-houses had also attained a certain importance there. Below this highest stratum of commerce, all those small industries were carried on, which then, as in our days, filled Italian streets with movement and noise. There were sellers of cakes, of sausages, of *frutti di mare*, who each, as Seneca tells us, advertised his wares in a particular tone and with a different cry.[2] They were called at Pompeii, people of the Forum (*forenses*) because they stand in public places. A curious picture shows us a cook, who has taken up his position in the open air, near his boiling saucepan, and is surrounded by a crowd of people who appear drawn by the savoury smell of his cooking. At the end of a stick he has a little copper cup, with which he draws from his saucepan what he sells to his customers.[3] This is a scene

---

[1] Some of these corporations, which did not include workmen of the same industry, but simply people who wished to live gaily together, bore strange names, like those which the Academicians of the Renaissance gave themselves. There is the society of sleepers (*dormientes*), that of late drinkers (*seribibi*), and even that of cut-purses (*furunculi*).

[2] Seneca, *Epist.* 56, 2.

[3] See Otto Jahn, *Ueber Darstell. des Handwerks*, etc., plate 3, No. 8.

that may be enjoyed every day in the markets of Naples.

The quarters where all these poor people dwelt have not yet been discovered. The smallest and most simply decorated houses yet excavated are not quite what we call houses of the poor. Perhaps some of them lived in those upper floors with terraces (*cœnacula cum pergulis*) which are occasionally mentioned in letting-bills. Unfortunately, only the ground floors of the Pompeian houses have been preserved; the rest has disappeared almost everywhere. Until the popular quarters are reached, the presence and habits of the lower orders can scarcely be revealed by anything except the places they delighted to frequent—here, as everywhere, the public-houses and taverns. At Pompeii there is no lack of these. At the entrance of the town are found hostelries intended for the peasants of the neighbourhood when they came to sell their wares or buy what they needed. Before the doors the pavement is lowered, in order that the cars may be able to enter the stables. It would have been inconvenient for them to circulate in the narrow streets of the town, where two vehicles would have found it difficult to pass abreast, so it was found more simple to leave them at the inn. These hostelries contain very small chambers, where travellers passed the night when obliged to prolong their stay. In some cases they have left their names upon the wall, with reflections which do not lack interest. It may well be thought that they are not great personages who content themselves with such poor lodgings. Among the number there are a Prætorian soldier on furlough, pantomimists come to give exhibitions, an inhabitant of Puteoli, who profits by the occasion to wish all kinds of prosperity to his native land (*Coloniæ Claudiæ Neronensi Puteolanæ feliciter!*), and a lover, who informs us that he has passed the night all alone, and that he has pined much for his charmer (*Vibius Restitutus hic solus dormivit et Urbanam suam desiderabat*).[1]

So here we are evidently in the company of very insignificant people. Those who haunted the taverns could scarcely have been more distinguished. At Pompeii the houses where hot drinks (*thermopolia*) were sold are very numerous. As with us, they are usually found at the places where there are most passers, and especially at the corners made by the junction of two streets. Before the door is placed a marble counter, with round openings, in which the vessels containing the drinks were placed, and little shelves on which glasses of different forms and sizes were to be ranged. This was for people in a hurry, who had not time to enter the shop, and desired to drink without stopping. If they were at leisure, and wished to be more at their ease, they went and sat down at the tables, in other rooms beyond the shops. Just such a shop as this was discovered a few years ago. It was decorated with curious paintings, well indicating both what kind of public frequented it, and that it was at the same time a gaming-house and a place of evil resort. One of these paintings shows the female servants of the tavern amusing themselves with the customers, pursuing them, embracing them, and inciting them to drink. Another

---

[1] *Corp. insc. lat.*, IV. 2146.

represents two bearded men with a gaming-table on their knees, and playing at dice. Both appear very animated. One seems triumphant at the good stroke he has just made, while the other shakes the dice in the box, in hopes of making a better stroke yet. In the next picture we have two players disputing. They load each other with gross abuse, reproduced by an inscription placed above their heads. At the noise the tavern keeper runs up, and with great politeness and in a respectful posture, begs them "to go and fight outside the door."

The various classes of Pompeian society of whom we have just made separate studies, did not always live apart from each other. There were frequent occupations and pleasures which brought them together. First of all, they were assembled by the care of public affairs and the election of magistrates. In these, all bore their parts, and after what would at first sight appear to be a very active fashion. While going over Pompeii, one's eyes are at every movement attracted by electoral posters; there is hardly a street in which one is not met with. In Paris the authorities take the trouble to have them torn down when the elections are over, but there are country towns where they remain for a long time upon the walls. This is what happened at Pompeii, and there are some several years behind-hand. They do not, like ours, contain professions of faith, in which the candidate sets forth his opinions. It is his neighbours, his friends, his protegés, who recommend him to the electors, affirming that he is a very good man and worthy of the functions to which he aspires. To read these numerous announcements, and observe the eagerness of so many persons to extol their candidates, one were tempted to think that the elections must have been very animated, and that the public offices of these small towns were competed for with ardour. This, doubtless, must often have been the case, but in certain election advertisements of Pompeii more of politeness is found than of politics. Some are the work of important personages who were candidates in preceding years, or soon will be such, and who wish either to requite a service or prepare themselves support. This exchange of good offices is displayed in quite a visible manner. A kind friend, who desires to gain someone over for the candidature which he is backing, tells him without ceremony: "Proculus, name Sabinus ædile; afterwards he will name thee thyself."[1] But oftener it is more humble folks—clients, people under an obligation, who would fain testify their gratitude and pay their debt after this clamorous fashion. The public offices cost so dearly that candidates could not always have been very numerous. It was, perhaps, because they had few competitors and their election was not doubtful, that they desired that it should at least appear to be the expression of the general will. The honour for them was less in the feebly-contested election itself, than in these manifestations which heightened its *éclat* and made its value. This is why the citizens thought themselves obliged to recommend themselves so vigorously to each other, although everybody was disposed to appoint them.

---

[1] *Corp. insc. lat.*, IV. 645: *Sabinum ædilem, Procule fac, et ille te faciet.*

When the impulse had seemed general, and opinion had declared itself noisily, the *duumvir* or the *ædile* was prouder of his success, and more disposed to acknowledge the good-will of his fellow-citizens by enormous liberalities.

Among these liberalities, those most pleasing to the people were the public games with which they were regaled. They had always been much loved in Rome, and were still more in favour in the country towns, where pleasures were fewer and life more monotonous. There were several kinds of them. First the scenic, for which two theatres were built at Pompeii, still in existence. I do not know whether many tragedies and comedies were acted in them, but certainly mimes must have been played. This little refined species of entertainment, which did not call for much literature, and was within the grasp of all, met with a good reception everywhere. Young people, especially, took pleasure in them, because, contrary to the general custom of the stage, the female parts were filled by women, these women being of easy morals, and an intrigue with a pretty actress a marvellous way of enlivening country life. Cicero said of one of his clients, whose youth had not been irreproachable: "He is accused of having carried off an actress; this is an amusement sanctioned by custom, especially in the *municipia*."[1] So pantomime was very much in vogue. It must have pleased at Pompeii, as elsewhere, and we know that Pylades, the great actor of Rome, came to give a few performances there, in the theatre erected by Holconius.

---

[1] Cicero, *Pro. Planc.*, 12.

who thus enriched the walls with their masterpieces. Children who were allowed to get hold of a piece of charcoal or chalk sketched a gladiator, as they now draw a soldier, and it is curious to remark that the manner in which those young hands proceed has not changed. The method is the same, and soldiers and gladiators resemble each other; the forehead and the nose always being represented by a line more or less straight, while two dots do duty for the eyes. However, some of these unfashioned sketches are not devoid of a certain comic meaning. I recommend to those who may have the plates of Father Garrucci before them, the arrogant attitude and bullying mien of Asteropoeus the Neronian, doubtless proud of his 106 victories (p. 11), and, above all, the stoutness of Achilles, surnamed the invincible (p. 12), whose sleek condition shows us that one did not always grow thin in this terrible trade.

Thus far, only the exterior life of the Pompeians has been in question, and it is that which, from the distance, is best seen. We follow them easily enough to the Forum and the theatre; it is less easy to penetrate into their homes. After the lapse of several centuries, it is always rather difficult to insinuate oneself into the private life of nations, to guess their intimate sentiments, their mutual relationship, their hates, their affections, their joys and their secret sorrows—all that the novel alone preserves and teaches to posterity. Yet we are much more fortunate at Pompeii than elsewhere. The abundance of the inscriptions discovered there enables us at least to divine what we cannot quite get to know, and enables us to sketch a few little broken-off romances which our fancy finishes and with which our curiosity is charmed.

Inscriptions were the only means of information and publicity then possessed, so they were very numerous in ancient cities. At Pompeii, three different kinds of them are found. Firstly, those graven on marble or stone, sometimes on the pediments of temples to inform us who built them, at others on the bases of statues, to tell us the names of the personages they represent, and the offices they filled. These inscriptions were designed to live as long as the monuments which bore them, and chance, in preserving them for us, has committed no indiscretion. Then there were those painted with a brush on the walls of houses or porticoes. These, much more curious than the former, played the part of the bills of our time. We have already spoken of those used to recommend candidates to the choice of the electors, or to announce the day and the programme of the spectacles. It is by means of them, also, that a landlord informed the public that he had an apartment to let for the Calends of July or the Ides of August, and that an innkeeper invited travellers to lodge with him, promising them a good dinner and all kinds of comforts (*omnia commoda præstantia*), and they are also used to advertise things stolen or lost, and to offer a decent reward to him who shall cause their recovery. "An urn of wine has disappeared from the shop; he who brings it back shall receive 65 sesterces (13 francs); if he brings the thief, he will be given double." The third species of inscriptions contains those which were simply traced in charcoal or graven with the point of a nail or a knife, either by the amorous who indulged in the pleasure of saluting their sweethearts by the way, or by some joker whom it pleases to inform us that he has the phlegm, or who unceremoniously treats as barbarians those who are so ill-bred as not to invite him to dinner, or by some wags, from whom we gather that Epaphra is a debauchee; that Suavis, the female wine merchant, is always athirst, and that Oppius is a thief. These *graffiti*, as they are called in Italy, were not made to come down to our days. The destruction of Pompeii has preserved them for us, and this is a great piece of good fortune. Truly, one little thinks how many things those scraps of boyish mischief, which garnish the walls where the police are tolerant, might teach posterity, if they got so far. It is without any doubt they which enable us to enter most deeply into the private life of the Pompeians.

In the *graffiti* of Pompeii a little of everything is found, even down to a washing-bill;[1] but what oftenest comes up is love. The goddess Venus was the patroness of the town; a patroness much respected, and invoked on all occasions. The people who ask you to vote for their candidate are careful to promise you the protection of Venus.[2] One of those impromptu artists of whom I have spoken, who chalked gladiators everywhere, finds no better means of protecting his drawing than to devote to the anger of the Pompeian Venus him who shall venture to touch it (*Abiat Venere Pompeiana iratam qui hoc*

---

[1] *Corp. insc. lat.*, IV. 1393.   [2] *Ib.*, 26.

læserit).[1] Lucian informs us that it was then customary to write declarations of love upon the walls. There are many of them at Pompeii, and their orthography being very varied, we may conclude from this that they were written by people belonging to different classes of society. Some, in order to celebrate their beloved one, are content to borrow lines from authors of renown, and, above all, from Propertius and Ovid. He was "the poet of light loves," and among young people none was more in vogue. At other times, the verses are drawn from writers now lost. A few even seem to be composed expressly for the occasion, and some of these are not badly turned for country verses. "May I die," says the happy lover, "if I would wish to be a god without thee!" (*Ah! peream sine te si deus esse velim!*) "Hither to me, enamoured ones!" says the irritated swain. "I mean to break Venus's ribs." (*Quisquis amat veniat, Veneri volo rumpere costas.*)[2] Some very pretty ones were found not long ago, certainly by a genius of the country. A lover is addressing the coachman who drives him. "Muleteer," he tells him, "if thou felt the fires of love, thou wouldst haste thee more to join thy adored one. Prithee, quicken thy pace; come, thou hast drunk well, take thy whip and wield it, bring me swiftly to Pompeii, where my dear amours await me!"[3] More often, the declarations are in prose. Sometimes it is the lover who gently supplicates. "My dear Sava,

---

[1] *Corp. insc. lat.*, 533. I change nothing of this barbarous Latin.
[2] *Corp. insc. lat.*, 1928 and 1824.
[3] *Bull. dell. instit. di corr. arch.*, 1877, Nov.

love me, I pray thee!" At others it is the loved one who answers: "Nonia salutes her friend Pagurus." These lovers have at times a delicacy of expression, and even something of refinement, which reminds us that we are in the age of Petronius. "My little doll, who art so pretty, he who belongs to thee entirely sends me to thee."[1] I prefer this more simple declaration, in which the heart appears to me to speak with greater frankness: "Methea, the player of Atellanæ, loves Chrestus with all her heart. May Venus favour them, and may they ever live in amity!"[2] And let us not forget this dismissal, given in due form to a wretched lover, and quite unanswerable: "Virgula, to her friend Tertius: thou art too ugly!" (*Virgula Tertio suo: indecens es!*)[3]

Of course I cannot quote all. I desire not to avail myself too freely of the permission granted to Latin to brave decency. If I dared put before my reader's eyes those libertine inscriptions which agree so well with the paintings of the secret museum, I should give him I fear, a very bad idea of the morality of the inhabitants of Pompeii, and, unhappily, this idea would be a true one. It was then generally pretended that morals were much better in the country than in Rome. Tacitus and Pliny everywhere delight to extol the decent and frugal life of the Italian *municipia*, and, to hear them, it would appear as though Rome were the rendezvous of all the vices, and virtue began just beyond the walls of Servius. I am very much afraid

---

[1] *Corp. insc. lat.*, IV. 1134.

[2] *Ib.*, 2457.   [3] *Ib.*, 1881.

that there must be in this opinion a little of that illusion which makes us think we should be far better off everywhere where we are not. At any rate, it was not true of the town we are just studying. Possibly virtue was not found in Rome, but it is certain that Pompeii was not the place to look for it. This charming town was situated in an enchanting country, where everything inclines to voluptuousness: "Where the velvet-like sheen of the fields, the tepid warmth of the air, the rounded outlines of the mountains, the soft windings of the rivers and the valleys, are so many seductions for the senses, which everything lulls to repose, and upon which nothing jars." It was near Naples, already called lazy Naples (*otiosa Neapolis*), which so well justifies the proverb that idleness is the mother of all the vices. It lay opposite to Baiæ—the most beautiful place in the world, but one of the most corrupted—to Baiæ, of which Martial says that the Penelopes who were so unfortunate as to venture thither became Helens.[1] All therefore combined to make of this country a sojourn dangerous to virtue, and inscriptions as well as monuments prove to us that Pompeii had not resisted these powerful seductions of climate and example.

We see what services we gather from these election-bills, these gay or serious advertisements, these jokes chalked up in passing by school-boys for fun, these simple or gross reflections of lovers or libertines. We possess the streets and houses of Pompeii, but empty and mute; the inscriptions and the *graffiti* seem to give

---

[1] Martial, I. 63.

us back the inhabitants. Pompeii comes to life again and is repeopled as we read them. We are no longer in the midst of ruins drawn with great difficulty from the ashes which had covered them during eighteen hundred years, but in a living town; and, as we pass through it, it teaches us, much better than books, what was done, what was thought, and how life passed in a country town in the first century of our era.

THE END.

www.ingramcontent.com/pod-product-compliance
Lightning Source LLC
Chambersburg PA
CBHW022103300426
44117CB00007B/573